CHINA'S WATER CRISIS
Zhongguo shui weiji

CHINA'S WATER CRISIS
Zhongguo shui weiji

MA JUN

Translated by Nancy Yang Liu and Lawrence R. Sullivan

AN INTERNATIONAL RIVERS NETWORK BOOK

EastBridge

Norwalk

Zhongguo shui weiji was originally published by
China Environmental Sciences Publishing House (Beijing) in late 1999.

EastBridge is a nonprofit publishing corporation,
chartered in the State of Connecticut and tax exempt under
section 501(c)(3) of the United States tax code.

EastBridge has received a generous multi-year grant
from the Henry Luce Foundation.

An International River Networks Book

Library of Congress Cataloging-in-Publication Data

Ma, Jun, 1968-
 [Zhongguo shui wei ji. English]
 China's water crisis / Ma Jun.
     p. cm. -- (Voices of Asia)
Includes index.
 ISBN 1-891936-27-1 (pbk. : alk. paper) -- ISBN 1-891936-28-X (cloth
:alk. paper)
 1. Water resources development--China. 2. Water--Pollution--
China.
I. Title. II. Series.
HD1698.C5 M3 2003
333.91'00951--dc21   2003011123

CONTENTS

Illustrations appear between Chapters Four and Five.

PREFACE

My work in journalism has given me many rewards but has also afforded me many opportunities to get a firsthand look at China's water problems, especially over the last decade. In those years I have seen the drying up of the Yellow River, devastating floods on the Yangtze, the rapid decline in northern China's water table, and the serious pollution problems of southern China's water. However, like so many other Chinese, I simply believed that those problems should and could be left to officials and engineers to take care of.

But after I had the chance to study the problem and interview people, I came to the shocking conclusion that they were simply trying to rob nature of the last drop of water to serve economic expansion. And while most people regarded the floods, dry spells, and sandstorms as some sort of evil force that demanded even larger engineering projects, I began to view them as nature's way of retaliating for man's reckless attempt to conquer and harness nature.

The Chinese edition of this book, which was published at the end of 1999, contained a detailed description of the water resource problems in all of China's seven major water basins: the Yellow River, Yangtze River, northwest, northeast, North China Plain, southeast, and southwest. (*Appendix 1 — Major Drainage Basins*) The conclusion was simple and stark: A water crisis loomed large in most parts of the country, one that would pose a major threat to social and economic development in the twenty-first century.

In general, the whole of northern China suffers from a scarcity of water, with the Yellow River being the most obvious example. The flow of this "mother river" began halting periodically in 1972, and in 1997 not a single drop of water reached the sea for a 330-day period. The once mighty river has by now become a small, filthy stream that cannot even flush much of its sediment into the sea.

But the Yellow is no worse than the situation on the North China Plain, most of whose 300 rivers are open sewers if they are not completely dry. Locals have continued to drill deeper to supply the needs of economic expansion, and the area's 60 billion cubic meters of nonrenewable ground water have shrunk by half in the past fifty years. Those supplies could be exhausted in some places within fifteen years.

The northwest's vast basin has fewer people but fewer water resources as well. Its rivers and lakes are vital to the fragile ecology, but most have either shrunk substantially in the past half century or simply dried up thanks to reclamation work and mining. Local people have become ecorefugees, and their work has left several hundred million people below the Great Wall exposed to the ravages of sand and dust. *(Appendix 2 – China's Deserts and Land Hit by Desertification)*

The country's southern half, unlike the north, is far more humid and has plenty of rain. There floods are the problem. And to deal with these, people have built mile upon mile of embankments and thousands of dams. But the devastating 1998 floods on the Yangtze showed that they were not the solution.

But then a rather strange new phenomenon appeared in the humid region: an odd sort of water shortage that was caused by pollution. To escape this, cities such as Suzhou and Wuxi and the megacity of Shanghai in the Yangtze delta, where every 300 meters one can see a river, have also turned to tapping the groundwater. This has lead to ground subsidence, which has actually worsened the effects of flooding. In the far south, in the Pearl River delta, where 13 percent of China's water is concentrated in a tiny area, prosperous urban centers such as Guangzhou (Canton) and Shenzhen regularly suffer water shortages.

People understand the dire consequences of the drying up of rivers and depletion of the aquifer, but many view water pollution as some sort of temporary growing pain that can be handled with a bit of investment. For the former, some people living on the North China Plain are busy drawing up plans to build two 1,200-kilometer canals to transfer water northward from the Yangtze. But in the case of the latter, the fact that several multibillion-dollar cleanup projects have failed has caused people to see, much to their horror, that pollution is a nightmare that could haunt them for years.

In many places I visited, I would no more than start asking locals about the water situation before finding myself surrounded by people willing and eager to talk. Many could recall how clean and beautiful the rivers had been fifty or even twenty years before, and they were all looking for an answer to the same question: When will our rivers flow like they used to? When will they be as clean as they were?

The questions were troubling. In China we have probably spent more time dealing with water problems than anywhere else on earth. In the past fifty years we built 85,000 dams of various sizes in addition to other facilities, and our water supply capacity was above 580 billion cubic meters per year. Urban water supplies increased a hundredfold. But both the arid north and humid south had worse water shortages. Why?

When I began to examine historical records, I found that China had been a whole lot different before humans began messing with it. Records showed that most hills and mountains had once been covered by primary forest and other vegetation that saved water and held the soil. There were far more lakes and wetlands along the lower reaches of major rivers that functioned as natural flood catchments.

Records also showed how these protective devices were methodically removed throughout the long course of civilization. I wrote of the slash-and-burn techniques used on the forests of the Yellow River's middle reaches 2,000 years ago and of how the same fate befell the Yangtze 300 years ago and the northeast and northwest 100 years ago.

But nothing could match the destruction of the past fifty years, when state-run logging firms cut down vital forests at the headwaters of major rivers, people left mountains and hills bare for hillside farming, and grassland and lakes kept being turned into even more farmland. Meanwhile, major rivers were blocked by huge dams, and many of these water ways just shriveled up or disappeared.

In the past two decades of modernization, fertilizers and pesticides have been applied recklessly for agricultural purposes, dangerous industrial chemicals have been dumped into water systems untreated, and cities have grown rapidly, paying little or no attention to sewage treatment and contaminating what little surface or subsurface water was left.

But don't get me wrong in my use of the title *China's Water Crisis*, with its sort of doomsday ring. In fact, I don't believe that we will run out of drinking water anytime soon, because for centuries China grew with limited resources. However, note the "anytime soon." China grew, of course, by robbing nature of its resources. As machines got better and money more plentiful, man could drill deeper and build larger dams or even be audacious enough to try to divert water from thousands of miles away — but there has been a limit.

Supporting a large population and rapid economic growth meant expending a lot of effort to increase water supplies. That unfortunately upset the ecological balance and caused resources to dwindle. When that conflict between increasing demand and dwindling resources reaches a certain level, there will be indeed a water crisis, so all that deeper drilling and reaching out for water thousands of miles away is actually a ticking time bomb.

This rather dark scenario has prompted some strategically minded people to propose international river projects, and plans are still afoot to build huge dams on the Yarlung Zangbo in Tibet (known in India as the Brahmaputra and in Bangladesh as the Jamuna) and the Nu (known in Burma as the Salween) and the Lancang (Mekong) to transfer water to arid parts of northwestern China. This could of course lead to environmental hazards and international disputes.

The Chinese government has in fact begun taking notice of the necessity to restore the ecological and environmental balance. Top leaders have placed emphasis on the more sustainable use of water resources, especially after witnessing the colossal floods of 1998, and there have been new policies to restore forest cover and lakes to their past glory. One might say that this has been the biggest concession ever made to the state of the nation's ecological health.

In reality, however, the fight between expansion and sustainability continues. More large dams have been planned for major rivers, even those that are dying. And the more rational pricing scheme that has been proposed to encourage people to save water hasn't been very popular. Instead, cities all across China have been preparing to get fresher water with the costly interbasin transfer project to solve their water shortage or water pollution problems. The biggest of these is the South-to-North Water Diversion Project, to bring water from the headwaters of the Yangtze to the Yellow River and Beijing.

History may in the future look back on this moment as a turning point in China's slow process of ecological deterioration, which began more than 5,000 years ago. If indeed we are witness to a radical policy shift—from blind reliance on water control projects in the name of "harnessing and conquering nature" to living in harmony with nature by balancing population, resources, and the environment—then this truly is a turning point. And it is the only way to stop China's water resource crisis.

Survival or death of water resources is the question. Allowing the water of our "mother river" to disappear right before our very eyes would be the most heinous crime we could commit. And, conversely, to rescue the dying rivers with our devotion and work would be our most glorious effort.

As the Tang Dynasty poet sang more than a thousand years ago: "All you and all your glory will perish one day, but the rivers will flow on forever."

CHINA'S WATER CRISIS
Zhongguo shui weiji

CHAPTER ONE

THE YELLOW RIVER

FLOATING YELLOW EARTH

There are hundreds of rivers in China, but there are only four major ones. Of these the Yellow River is the most prominent.
—*River Notes, History of the Western Han* (206 B.C.–A.D. 8)

THE YELLOW RIVER IS THE FIFTH LONGEST RIVER in the world, traversing 5,464 kilometers of the vast territory of China. It arises from glacial springs on the Qinghai-Tibetan plateau and drops 5,000 meters in its eastward course through a series of gorges, bringing much-needed water to the arid north.

This is no ordinary river. It has long been revered by the Chinese, for whom the color yellow holds a special meaning: the yellow earth, the Yellow Emperor, and even the yellow-colored skin of the people. These are all cultural signs that have caused people throughout history to assign an almost divine status to the murky waters of the river. *(Appendix 3 – Yellow River Drainage Basin)*

"Mother river" is how the Chinese have traditionally described it. But, over the centuries the Yellow River has come to be known not only as China's savior, but as its sorrow as well. It cuts through the deep Loess Plateau of northern China, a land of parched, cracked earth largely void of trees and grass, and has often left its imprint on the nation with a startling impetuosity and ferocity.

Many have wondered how such a harsh, unforgiving, and even hostile environment as that of the Loess Plateau could be the cradle of Chinese civilization. What people might fail to understand is that up until about 6,000 years ago, there was green everywhere, with forests covering some 59 percent of the land.

What Chinese refer to as their 5,000 years of civilization could in fact almost be called 5,000 years of ecological degradation. Generation upon generation of Chinese started first with the trees and then went on to denude the area of its grass and bushes to turn it into farmland. Once the forests on the plateau began to disappear, the loose, coarse soil and sediment began to have serious erosion problems. In no time at all, the face of the environment changed dramatically.

Today, the forest cover on the Loess Plateau has dropped to only 6 percent and the plateau, with its lunar surface, has come to live up in the starkest way to its name, because loess is all it is: wind-borne dust of minute yellowish gray grains packed into a friable soil varying in depth from a few feet to over 250. *(Appendix 4 – Loess Earth and Rock in China)*

The appalling soil erosion continued unchecked for several thousand years because the loess here is an astonishing 100 to 300 meters thick. But the once smooth landscape is now a grotesquely disfigured mess, scarred with some 30,000 gullies and ravines up to a kilometer long and a vast number of cracks and fissures, large and small.

These have become the primary source of the pervasive yellow precipitate in the river. It is believed to account for 90 percent of the 1.6 billion tons of annual sediment the river receives. This is three times the amount of sediment of the Nile, the Mississippi, and the Amazon combined.

So silted up is the river that one-half of a bowl of its water will be sediment. If the river's sediment were used to build a berm 2 meters by 1 meter, it would stretch from the earth to the moon. Small wonder, then, that the river is now referred to simply as "floating yellow earth."

"Waters of the Yellow River Come Flowing from the Sky"

There used to be no embankments along the Yellow River but at that time there were numerous lakes and ponds in low-lying places to hold the overflow during the rainy season so we did not have floods.
　　　　—River Notes, History of the Western Han (206 B.C.–A.D. 8)

During the flowering of the Tang Dynasty (A.D. 618–907) the poet Li Bai had this to say about the Yellow River: "Do you not see the waters of the Yellow River come flowing from the sky? The swift stream flows into the sea never to return."

The poet's words were prophetic, but the future was stranger than his description. The waters of the Yellow River do indeed flow well above the heads of the people: For the last 800 kilometers of its journey across the North China Plain, the river is normally 5 meters above the surrounding land.

In the city of Xinxiang, in Henan province in east-central China, the main channel is a full 20 meters above the surrounding countryside. One might say that the Yellow River is floating on air, magically suspended.

This is not the result of a long, slow, natural process. Quite the contrary; it is the result of man's tampering—a deliberate attempt to "conquer and harness nature." Chinese attempts to take serious action against floods started long ago and continued for more than two millennia.

The Chinese accounts start in the mists of legend. Figures became legendary for their heroic attempts to divert the raging Yellow River waters of 5,000 years ago. By 2,600 years ago, the Chinese already had some fairly large dikes situated along the lower reaches of the Yellow River. There was already ecological destruction along the upper reaches, making the water of the badly silted river murkier.

As the bed of the river gradually rose, the dikes were built higher. A part of the river was already rising above the surrounding countryside more than 2,000 years ago. This greatly increased the potential flood damage. Some sages who perceived the danger suggested that people be moved out of the low-lying land and that the river be allowed to spill over naturally and run its course. But China already had 60 million people living in its more populated parts, and the demand for food and clothing made even these sensible concessions unacceptable. China's rulers chose instead to build the dikes higher.

The enormous prosperity of the Tang Dynasty paradoxically resulted in even greater destruction. By the latter part of that imperial era, floods were becoming a much more serious threat than they had been previously.

During the 300 years of the Song Dynasty (A.D. 960–1279), which followed the Tang, there are records of as many as forty separate incidents of major embankment collapses. The increase in the amount of soil erosion led to a vicious cycle of more sediment being deposited in the river, causing the bed to rise every year, producing more dangerous floods.

In desperation, people dug additional stream branches and tributaries alongside the river. But these soon silted up as well, until the answer became obvious, even to the most obtuse: The harnessing of the Yellow River had to begin with the sedimentation problem.

By the middle of the Ming Dynasty (mid-sixteenth century), a minister, Pan Jixun, spent twenty-seven years looking at different ways to deal with the sedimentation problem. He finally came up with what seemed to be a brilliant approach: Make the riverbed narrower through the use of embankments, thereby increasing the speed of the water, and the amount of sediment washed away would increase substantially.

For several generations this remained the guiding principle of river control, since it was believed that more than 1.2 billion tons of sediment was washed annually down the river valley to the delta and eventually into the sea. And it has continued to be applied, even up to the present. Pan's idea was followed in the case of the Three Gate Gorge Dam (Sanmenxia), which was completed in the 1960s, and more recently in the Xiaolangdi water control project just north of the city of Luoyang.

But the narrow channel still could not take care of all the sediment being washed into the river, nor could it solve the problem of the silt accumulating in the riverbed. What is more, the narrowing of the channel meant less space for the silt to accumulate on the bottom, so the riverbed began to rise dramatically. This was when the Yellow River became suspended even more precipitously, posing a danger to the surrounding land.

By the Qing Dynasty (1644–1911), the scale of these river control efforts became even greater — they were unprecedented, in fact — but the problem refused to go away. The tragic effects could be as local as they were widespread. Several officials who had been appointed to oversee river management ended up committing suicide after they were disgraced when Yellow River floodwaters defied them and devastated large parts of northern China.

Those attempts at river management had evolved into a vicious, often deadly cycle: Embankments and levees were reinforced; sediment accumulated and the bed of the river rose; that meant further reinforcement of embankments, which served to increase the accumulation of sediment. The higher the embankments and dikes became, the more dangerous the situation in the surrounding countryside became. The Yellow River, in a relatively short period of time, became suspended in space—a sword of Damocles hanging as if by a thread over the heads of the people.

The vicious cycle continued until 1949, when the People's Republic was established. The new leaders felt that they were a different breed, equipped with modern machinery and advanced ideology. The new China's response was simple: just get even tougher with the environment. For five decades, all the government's energy went into intensifying efforts to "harness the river and control nature," and, the Yellow River was the primary object of that wrath.

"LET ALL MOUNTAINS AND RIVERS MAKE WAY FOR US"

The completion of the Three Gate Gorge reservoir will fulfill a thousand-year dream of the Chinese nation, that is to see the Yellow River become clear.
—Vice Premier Deng Zihui, 1956

THE LEADERS OF THE NEW PEOPLE'S REPUBLIC OF CHINA, in ordering an unprecedented amount of embankment work along the Yellow River, reflected their anxiety about putting an end to the centuries-old problem of flooding once and for all and their impetuosity about doing everything they could to force the river to serve economic development ends. When their anxiety and impetuosity merged, it meant that they became obsessed with reservoirs as a way to control the floods and provide water for both industrial and agricultural use.

In 1952 Communist Party chairman Mao Zedong, as the newly installed leader of China, made his first full inspection of the country, which included a visit to the Yellow River. When he arrived at Mang Hill, just north of the city of Zhengzhou, in Henan province, he climbed up onto the eastern side of an earthen dam and, looking at the rapids, was said to have murmured: "What's to be done if the waters of the Yellow River rise sky high?"

One of the people accompanying Chairman Mao was Wang Huayun, the director of the Yellow River Conservation Commission, who, after taking in Mao's query, spoke directly to the chairman in a voice everyone could hear: "The only thing to do is to build a huge dam. Small-scale dams won't amount to a hill of beans."

Mao was especially fond of such grandiose ideas since they coincided with his general philosophy that humans should "conquer and harness nature." A year later he made another visit to the area. After studying the engineering charts for the Three Gate Gorge Dam in Henan province, Mao asked just how long it would take after the construction work was finished for the dam to begin operating.

Wang Huayun didn't hesitate: "If the 30 power stations planned for the main stream are completed, total reservoir capacity should be between 200 and 300 billion cubic meters. Under those conditions, if little or even no effort is put into controlling soil erosion and not even one reservoir built on the tributaries, the big dams themselves can function for well over 300 years."

Encouraged by this rosy prediction, Mao was determined to move ahead with plans to build a really big dam. There were still four options, but Wang Huayun tried his level best to convince Mao that the grandest was the best. He believed that a huge dam that cut off the Yellow River could impound water to stop sedimentation.

By 1953 China was following the Soviet Union's lead in developing its own five-year plan (1953–57) for the nation's economy. At the same time, discussions with the Soviets on 156 projects they had promised as the focus of an aid package to China were intensifying. China's Water Resources Ministry (Shuili bu) and the Yellow River Hydrology Committee (Huanghe shuili weiyuanhui) jointly requested that the Yellow River project be included as a major recipient of Soviet aid.

In January 1954 China received a delegation of Soviet experts. They agreed with the Chinese that the Three Gate site, the last gorge before the Yellow River entered the North China Plain, was the optimal site for a big dam. Just like that, Wang

Huayun's novel if not downright bizarre idea became national policy—almost in the blink of an eye, one might say.

By 1955 the grandiose Yellow River plan had forty-six separate hydroelectric power projects (the "staircase plan"), from the Longyang Gorge (Longyangxia) in Qinghai province to further downstream, as its centerpiece. Their purpose was to gather water to produce electricity and for use in irrigation. That number, however, was ultimately reduced to twenty-nine. They included seven key projects, and the Three Gate Gorge project became the litmus test.

The Three Gate Gorge Dam is at the juncture of Shaanxi, Shanxi, and Henan provinces. It is less than 400 meters wide and is a marvel. Smack dab in the middle of the river there are two rocky islands that split the flow into three sets of rapids—the "three gates"—each with its own name: Gate of Ghosts, Gate of the Gods, and Gate of Man.

The design work for the large structure was entrusted to the Soviets, who said that a dam about 100 meters tall would be right. Thrusting their chests out, they assured their Chinese counterparts that that particular design would allow enough sediment to flow through the dam and not cause major accumulation problems in the reservoir.

The work began in April 1957, and by September 1960 the main structure was complete.[1] The Sino-Soviet romance went sour at the end of the '50s, and many large-scale projects, especially those supported by the Soviets, ground to a halt. The Soviets were sent packing, and their assistance was withdrawn after 1960. But that did not include the Three Gate Gorge Dam. Work on the dam continued at a frenetic pace, and 870,000 people were forced to move elsewhere to accommodate the project. That included several hundred thousand who were packed off to the remote and barren Ningxia-Hui Autonomous Region. But the Chinese and Soviet planners had promised that the Three Gate Gorge Dam would mean a "clear Yellow River" downstream because vast amounts of silt would accumulate behind the structure, so there were very few complaints about the project to be heard.

By 1960 the reservoir began to impound water. The dam, built out of 2.1 million cubic meters of concrete, stood gloriously tall astride the ancient river, cutting it in two and producing a lake that was larger even than Lake Tai, China's third-largest freshwater lake.

But almost before the applause had died down, a mere two years after the dam began backing up water and before even one kilowatt-hour of electricity was generated, the giant reservoir had already accumulated 1.91 billion tons of sediment. Only 1.12 billion tons of the stuff had been flushed downstream.

Not only was there a "big belly" of sedimentation in the reservoir, there was also an oversized "sand tail" in the upper reaches. The bed of the river upstream began to rise dramatically, and the mouth of a nearby tributary, the Wei River, was blocked by sediment buildup so rapid that it posed a threat to the good citizens of the ancient city of Xi'an and the fertile Guanzhong Plains.

In December 1964 Premier Zhou Enlai convened a conference to discuss a variety of issues related to the Yellow River. At the meeting he proclaimed that priority was to be given to "protecting Xi'an and the lower reaches" and that the original policy of impounding water to block sediment was to be abandoned. An anxious Mao even went so far as to suggest that the dam be destroyed by aircraft with bombs if no other option was available in the effort to save Xi'an.

After almost a decade of hard reconstruction and alteration work, the Three Gate Gorge Dam was saved. Still, it was not until the end of 1973, a long thirteen years after the dam was completed, that it began generating some electricity. However, the reservoir, at only 1.8 billion cubic meters of water, was considerably smaller than the grandiose 35.4 billion cubic meters that was originally intended, so its ability to generate electricity and to provide irrigation was seriously compromised.

But that immense setback did not seem to create much of an impression on the Chinese leaders, nor did they seem to ponder the negative impact of huge dams. Instead, opting for a more upbeat and positive outlook, they thought they had found a way to operate a dam even on the murkiest of rivers.[2]

And for several decades, that blueprint from the 1950s became an even larger reality as large hydrology stations popped up on the main stream of the Yellow River. There were 3,300 large and small reservoirs on the river and its many tributaries, giving a total reservoir capacity of 53 billion cubic meters.

At the turn of the century, the Wanjiazhai water diversion project began. Meanwhile, the process of adding reinforced concrete to the Lijia Gorge dam was completed. It was at this point that the work at Xiaolangdi, said by some to be one of the most difficult projects in the world, sped up. These three projects would bring total reservoir capacity to 66 billion cubic meters.

There was one major problem, however: The Yellow River's entire natural flow is less than 56 billion cubic meters annually on average. To make matters worse, when the Daliushu, Zikou, and Longmen dams are completed sometime in the foreseeable future, total reservoir capacity on the main stream and the tributaries will be a colossal 98 billion cubic meters. That would be 2.5 times the annual flow of the past three decades.

The Yellow River is today dotted with dams that were meant to perform several functions: control floods, generate electricity, and provide irrigation, transportation, and fish. But before the people could finally claim victory and crow about conquering the Yellow River, it pulled a horrific joke on them.

NOT ONLY THE SORROW OF THE RIVER GOD

When the autumn floods came, hundreds of tributaries contributed a large volume of water to the Yellow River. The river became so wide that people could not tell a horse from a cow on the other side. Riding the torrential stream down toward the sea, Hebo, the God of the Yellow River, thought there was no greater force than his river
    —*Autumn Flood*, Zhuang Zi (369–286 B.C.)

IN THE SUMMER OF 1972, staff at the Lijin Hydrometric Station, close to the mouth of the Yellow River, were shocked to see the bed of their river lying exposed to the air, cracked, with not a drop of water to run into the Bohai Gulf. No one could recall ever having heard of, let alone seen, the revered Yellow River not connected to the sea.

That strange phenomenon was thought at the time to be connected to the unusual drought that was bedeviling the area, so they didn't pay it much heed. Then the dry spells on the Yellow River began coming more frequently—six times in the 1970s, seven in the 1980s, and seven during the first eight years of the 1990s. People began to see that this was no one-time event.[3]

What was more serious was the fact that the dry area began to expand and the cracks gradually meandered off toward the central part of the North China Plain. During the 1970s the average length of the dry area was 130 kilometers; in the 1980s it grew to 150 kilometers, and in the 1990s it was 300 kilometers long. By 1995 it had grown to a startling 700 kilometers in length.

Not only were the parched and cracked areas expanding, the periods of time when the river was completely dry or nearly so began to lengthen significantly.[4] The period in 1997 was 226 days long; for the 136-kilometer stretch from the Lijin Hydrometric Station to the mouth of the Yellow River it was 330 days. In view of the fact that a year has only 365 days, people had good reason to worry about the possibility that the river could one day dry up not just for a number of days, but for months or even years.

In fact, proposals that the Yellow River be considered a seasonal phenomenon began appearing some years ago, but they were quickly dismissed as fantasy. Then this strange phenomenon occurred. The Yellow River had indeed become a seasonal attraction, with significant dry periods in much of the area it drained. Not only that, but it began to appear as if the river might actually stop flowing into the sea altogether and become an inland body of water.

After having sung and danced to the "conquer and harness nature" tune for so many years, man, it turned out, had not defeated nature; he had defeated himself. The joke was on him: the complete death of the Yellow River. Its floods might not bother him anymore, but that was because its water did not even flow into the sea.

Now some people were screaming, "Protect the Yellow River!" while others betrayed little emotion and could not care less. But still, for many, the reaction was one of frustration and befuddlement, with absolutely no sense of what could be done.

The Yellow River, it turns out, is far too complex an ecosystem for a handful of people to completely understand, let alone manage or control. The solution, it seems, had always been to sacrifice the river in the name of mankind's interests.

Some scientists suggested that people stop fooling around and just go ahead and convert it into an inland river. They were of the opinion that, in view of the enormous water needs of northern China, not a single drop of the river's water should be allowed to flow into the sea.

Then there were the officials who did not want to address it in such a straightforward fashion. But the decisions they made to build more dams along the river would have the same effect in the end.

Some environmentalists have argued that the river should be restored to its natural state, but this back-to-nature idea was immediately dismissed as being "unrealistic," and the environmentalists were left talking to themselves.

Is the Yellow River water that does escape to the sea in fact a waste, as the officials and hydrologists claim? Obviously it's not. But why?

First and foremost, a person needs to bear in mind that no matter where the Yellow River's waters end up, sand and sediment follow. To flush even 1.2 billion tons of its 1.6 billion tons of sediment into the sea annually requires something on the order of 21 billion cubic meters of water.

And yet, by the 1990s, the average annual amount reaching the sea had been reduced to 10 billion cubic meters, and that was expected to drop significantly in the new century. This results in an increase in the amount of sediment accumulating in the lower reaches, and a dramatic increase in the amount of flooding.

In addition, desertification of the lower reaches has been a serious side effect. Over the years, studies have shown, only 43 percent of the coarse sediment flowing into the riverbed reaches the delta below the Lijin Hydrometric Station.[5] That means 57 percent stays behind in the riverbed.

Once the delta begins to dry up, the sediment left behind in the riverbed will dry up rapidly, leaving a very fine residue of earth that can easily be borne aloft by the harsh winter winds and deposited elsewhere in large, shifting sand dunes, posing a serious threat to the prosperous North China Plain.[6]

If the river dries up, it strangles the source of life of two provinces downstream, Henan and Shandong. Then the battle for water is on.

Henan is in east-central China and has 91 million people, making it the most densely populated of China's provinces. It is also a major agricultural region. Its annual per capita water consumption is only 467 cubic meters, less than a twentieth of the world's average. A hectare of land there gets only 6,225 cubic meters of water annually, or 26 percent of the average in China.[7]

Minor droughts hit the province every three years and major ones every five. Perennial water shortages have hampered both industrial and agricultural development immensely. As the Yellow River slowly dries up, Henan's cities, so long dependant on the river's water, will begin to undergo large-scale dehydration.[8]

The situation is no less dire in Shandong, the last province drained by the Yellow River before it spills into the Bohai Gulf. Its population accounts for 7.2 percent of the nation's and its arable land for 7.3 percent, but its water resources amount to a mere 1.2 percent of the national total. In recent years, Shandong's drought conditions have grown worse, with provincewide droughts occurring almost every year.[9]

The Yellow River runs for 617 kilometers across the province and is its lifeline. When the river has less water, just about every city along it suffers from shortages. To try to solve this problem, in the 1980s many of these cities began digging wells to tap into the aquifers. More than 10 billion cubic meters of water is now extracted from the ground annually. The result has been the rapid expansion of underground water funnels and cones of depression. They now cover 16,500 square kilometers and in some cities go more than 70 meters deep.[10]

Industries and cities are given greater access to water supplies. Rural parts of Shandong, by comparison, have suffered much more during dry periods. In 1997 alone there were 1.3 million people without adequate drinking water because of a dry Yellow River. Needless to say, agricultural production was seriously affected.[11]

The people of these two provinces have a good reason to worry about their future or even their survival.

YELLOW RIVER FLOWING TO THE SEA

The sun above the mountain glows, the Yellow River seaward flows.
　　　　—Wang Zhihuan, Tang Dynasty (A.D. 618–907)

NO RIVER ON EARTH CARRIES MORE SEDIMENT than the Yellow River, and no river creates as much land as the Yellow River. If it were to dry up, the delta region and its unique biodiversity could be destroyed, as was the case with the Aswan Dam and the Nile Delta.

The Yellow River's estuary in Shandong formed one of China's three large deltas. More than a billion tons of sediment has washed down to the river's mouth annually in the past several centuries, and 27 percent of that washes on out into the sea; the rest is deposited in the delta. Each year the delta pushes about 20 square kilometers farther out into the sea, one of the world's major landfill projects.

The delta covers 17,670 square kilometers of ground and has two cities, Dongying and Binzhou, and twelve counties.[12] It is also an important grain, cotton, oil, salt, chemical, and petroleum production center.

But as the dry periods grow longer, the river's capacity to create more land will be reduced, so that in the not too distant future, the province's coastline will start to recede. Since the 1970s, the amount of sediment flowing into the sea has been lessened considerably.[13]

The desertification of the delta area is an additional problem. Sediment has covered part of the estuary in a very short period of time, and grasses and other flora are becoming stunted and are easily destroyed. In addition, close to the sea, the water table is not very deep and the subsurface soil in the delta has a high saline content and is infused with mineral-laden saltwater, which drives the salt up toward the surface.

The surface of the area is very uneven and tides wash across it, which increases the salt content at the surface. The one force that counteracted this seawater intrusion in the past was the heavy sediment pouring forth from the mouth of the Yellow River.

This counteractive force is not around for much of the year now, and there has been a decline in the replenishment that fresh water and sediment bring. If the river were to dry up completely, the delta's downfall would not be far behind.

With this threat staring them in the face, people have devoted a considerable amount of resources and manpower to trying to hang on to every last drop of water from the upper reaches of the river—by building more reservoirs.[14] But as the river dries up more in its upper reaches, these expensive hydroelectric and irrigation projects end up as nothing other than white elephants.

Along with the reduction in the volume of water has come a decline in the water's quality. Even when the river is flowing, its water has become virtually useless in many places along the lower reaches. A 1994 study indicated that potable water was available in only 31 percent of the entire drainage area of the Yellow River. Most of that was in undeveloped, pristine areas to the west of Longyang Gorge in Qinghai province. For 4,507 kilometers of its entire length, the river has nonpotable water with toxicity levels at category IV and V or above. The pollution has also become more serious in recent years.[15]

People living in the delta have been forced to tap into the aquifer. There is a problem with this strategy, however, which is that iodine and fluorine levels of the underground water are often well below or well above acceptable levels. Not surprisingly, there have been high incidences of thyroid problems and forms of fluorine poisoning reported in the region.[16]

The environment and its fish, wild herbs, and endangered bird species have been subjected to greater dangers because they rely so heavily on the tiny organisms and other nutrients that are supposed to be a part of the sediment carried down by the Yellow River waters.[17] In recent years, when the volume of water reaching the mouth of the river was dropping, there were large reductions in the amount of these microscopic organisms and nutrients. This was a disaster that had already arrived.

The Yellow River flowed for thousands of years from its headwaters all the way to the delta, through nine provinces and autonomous regions. How had it arrived at this deplorable condition? What role have the various places along its course played in this decline? Who is ultimately responsible for the desiccation of an entire river, China's "mother river," and the ensuing crisis?

To answer this, we need to go back to the very beginning.

QINGHAI: THE LONG-HIDDEN SOURCE

Few have ever seen the remote ground in Qinghai.
—Du Fu, Tang Dynasty (A.D. 618–907)

EXACTLY WHERE THE YELLOW RIVER ORIGINATES has long been a topic of speculation and the subject of a considerable amount of research. For thousands of years, people living near the river have asked this question, and many expeditions have set out in search of the true source, out somewhere on the Qinghai-Tibetan Plateau. A wide range of people—some Han, some from ethnic groups, even some foreigners—have crossed treacherous lands trying to establish once and for all just where it comes from.

Even the chairman himself, Mao Zedong, once started preparing for just such an expedition on horseback. That plan was canceled, however, when relations with the Soviet Union began deteriorating even more in the early 1960s and Mao began to fear that there could be war. There have been reports that his never having taken that trek was something Mao regretted for the rest of his life.

In 1978 yet another expedition of the Yellow River Conservation Commission's River Source Research Party set off in an attempt to discover the river's headwaters. After scrutinizing and analyzing three tributaries—the Zha, the Yokutsunglieh (Yueguzonglie), and the Kari—that feed into the Xingsu Sea (Constellation Sea, a wetland with thousands of small lakes), the scientists decided that the most powerful of the three, the Kari, was the main source of the Yellow River.

The three streams meet at the Xingsu Sea, after which they feed two lakes, Gyaring and Ngoring, the plateau's largest bodies of water. The Ngoring is 30 meters deep and is revered by Tibetans living in those parts for its 10.7 billion cubic meters of crystal-clear water, which makes it China's fifth-largest freshwater lake.

The expedition completed its work and submitted a report. The area was soon forgotten by most people—until, that is, the Yellow River began drying up and making people think more seriously about its fate and its possible impact on the nation.

In Qinghai province, the Yellow River is approximately 1,960 kilometers long, about a third of its entire length, but it is fed from various sources that provide 28.5 billion cubic meters of water, or 49.2 percent of its volume. The province is the source of the Yangtze and Mekong rivers as well. But despite Qinghai's remote beauty, it is on the edge of an unfolding tragedy of immense proportions.

Of the province's 720,000 square kilometers of land (slightly larger than the U.S. state of Texas), 334,000 square kilometers suffer from serious soil erosion problems. Because of an excessive amount of logging, overgrazing, and land reclamation, a large part of its grassland has become desert. In 1949 Qinghai had 5.3 million hectares of desert, but by the late 1990s that figure had grown to 14.47 million.

Of those parts of the province drained by the Yellow River, serious soil erosion affects 75,000 square kilometers. In fact, of all the provinces drained by the Yellow River, Qinghai ranks number one in soil erosion, with large swaths of what was once grassland now nothing more than parched, cracked earth scored by countless gullies, hardly different from the Loess Plateau of Shaanxi and Shanxi to the east. The amount of sedimentation Qinghai contributes to the Yellow River annually is now about 88 million tons.

Qinghai's Maduo county is closest to the river's source. Maduo used to have more than 4,000 lakes, and as late as the early 1980s, vast herds of horses and sheep grazed there, giving the county a per capita income that was one of China's highest.

But Maduo has a problem: 70 percent of its grasslands have virtually disappeared. They're now barren, unfertile land with worthless dirt. Lakes and rivers have dried up by the hundreds if not thousands, and the water table is so low that many wells are useless. Desertification is growing by 20 percent annually, forcing some 10,000 people to become "ecological refugees."[18]

Dari county lies just to the south of Maduo, two counties away from the headwaters of the Yellow River. In the early 1970s, Dari had over 79,000 head of livestock, making it Qinghai's richest county in animal husbandry. But here too there has been significant deterioration, especially in the grass that herds used, because of overgrazing. Many areas by now have deteriorated seriously. On 92 percent of the county's grassland, the deterioration rate is 17 percent. As in Maduo's case, more than a thousand herdsmen have had to move to other areas. Some have ended up as beggars, while others have had to lease grassland just to survive.

Further downstream, just outside Qinghai and into Gansu and Sichuan provinces, the lush marshland has also dried up and declined so seriously in recent years that the contribution to the Yellow River has declined notably.[19]

The impact is felt not only on the Yellow River but on tributaries such as the Huang River in Qinghai. In areas drained by the Yellow River, forest coverage has dropped to as low as 7 percent. An area that was once lush and green has been taken over by yellowish brown scrub with little growth and almost no ability to stop soil erosion. The result has been that the clear water of tributaries like the Huang has become turgid, with a sediment content now as high as 7.75 kilograms per cubic meter.

The effects are clear: The Yellow River's water volume has dropped alarmingly. Every year since the mid- to late 1980s there has been a gradual reduction in water flow from the upper reaches. This accelerated markedly in the 1990s.[20] This, more than anything else, was the reason for the increasing number of dry periods in the lower reaches of the river.

The attempts to confront this crisis in Qinghai generally followed old policy guidelines—build huge water control projects. The Longyang Gorge Dam, which was completed in the late 1980s in Qinghai's Gonghe Basin, was designed as the first rung on a "ladder" of dams to be built along the entire length of the Yellow River. It is 178 meters high and has a storage capacity of 24.7 billion cubic meters, making it China's second-largest reservoir. But the recent environmental deterioration around the source of the Yellow River has put the Longyang Dam at risk.

Fifteen years ago, the dam was edged with grass that was taller than the sheep grazing in it. In the dam's vicinity now there is nothing but sand, deposited there by the harsh winds that punish the region. Sand dunes shift closer to the dam, and the reservoir suffers from nearby landslides.[21]

Because of annual reductions in the amount of water at the upper reaches of the Yellow River, reaching full capacity seems impossible. In recent years, the amount of water impounded has been about 9 billion cubic meters or less, or 15.7 billion cubic meters short of the designed capacity. The electricity generated has only been a quarter of its generating capacity.

The headwaters of the Yellow River are no different from the lower reaches and have dried up periodically since 1996. Although the immediate effect may be felt only by several thousand local nomads, it is still an ecological alarm.

The situation confronting us is clear: The river dries up in the lower reaches and takes away much-needed supplies from the people. The river dries up in the upper reaches and takes away the entire river.

A HIKE THROUGH GANSU

No place in the world is richer than the central steppe in Gansu.
        —*Records of the Grand Historian* (90 B.C.), Sima Qian

THE YELLOW RIVER FLOWS EASTWARD for 1,960 kilometers through Qinghai province and on through the narrow corridor that is Gansu province. Its 913-kilometer journey through this province is not a happy one. In Gansu, the waters of China's "mother river," already buff-colored, get muddier from the endless sand and mud deposits and from the other effects of long periods of drought.

The older Chinese name for Gansu was Long (pronounced "loong"), which means "steppe" or "high plains." The region through which the Yellow River passes is predominantly one of deep yellow dirt, a part of the Loess Plateau. To the west it gradually joins the Qinghai-Tibetan Plateau, and to the north the Inner Mongolian steppe, where the green of China comes to an abrupt end.

Around 30 percent of Gansu was covered with forest for much of its history. In the central steppe region through which the Yellow River passes, the percentage was much higher. With abundant woods, grass, and fertile land, the Yellow River valley of Gansu was traditionally thought of as something like a paradise where herdsmen could raise strong horses and fat sheep.

This natural beneficence was a magnet for herdsmen, who migrated to the area in droves beginning as far back as the Qin Dynasty (221–209 B.C.) and the Han (206 B.C.–A.D. 220). The central and southern parts saw their valuable grasslands quickly overrun and turned into farmland. As a result, large swaths of fertile grassland and forests were virtually destroyed. With the dramatic loss of vegetation, the Loess Plateau rapidly fell prey to soil erosion.[22]

This chaotic pattern of destruction continued until the twentieth century and the Nationalist period (1912–49). By 1949 the province's forest cover had dwindled to only 6 percent, and the central steppe sank deeper into the morass of environmental degradation.[23] The once-rich land with lush foliage had become a place suited only for beggars.

Curiously enough, although survival in this harsh environment became more difficult, the population grew—from 25 million in 1949 to over 60 million by 1990. Thus began a vicious cycle. Population growth led to greater poverty, which led to increased timber losses and poorer conditions for people, who in turn saw large families as the only way out of their pitiful condition. This more than halved the access to arable land, forcing people to claim more hillside land and leading to a further decrease in the forest cover. To no one's great surprise, productivity in the province's eighteen counties dropped precipitously.[24]

This destruction of trees and vegetation, combined with other factors, led to climatic changes in the central steppe. Meteorologists reported that for the decade and a half leading up to 1950, the weather remained relatively constant 41 percent of the time. But from 1950 to the late 1970s that figure changed to 14.8 percent, meaning fewer years with mild breezes, ample rainfall, and other benign weather conditions.

Floods became the norm, alternating with severe drought. Today it is not unusual for the region to go several months without adequate precipitation, denying the people the water they need for irrigation and household use.

During the drought season, some areas have become so dry that even the lowly sparrows got almost nothing to eat. Since there were no new water resources for them to seek, the birds died off. Humans want to continue their lives no matter how miserable they are, so water somehow needs to be found, whatever the method. If a few dark clouds appeared on the horizon after a prolonged period of drought, the people went off with their pots and pans in search of even a drop of rain. The same pattern could be seen in mountainous areas, where people hoped to capture even a little bit of the runoff tumbling down cliffs. In these ordeals, patience was not a virtue; it was the only way to exist. They knew that their previous drop of water could be their last.

It is easy to see why the "121 projects for collecting rainwater" (a form of dry-land cultivation used in parts of the world since ancient times) came to be known as a lifesaver. The idea was that within a 100-meter radius of every household, people would go about the task of collecting rainwater by digging ditches, cisterns, and other catch basins to get enough water for humans and animals.[25]

But before they resorted to this last-ditch effort, Gansu had already gone all out building huge reservoirs on the Yellow River. For decades the province had put nearly 80 percent of the government spending into dams. In the relatively short 913 kilometers of the river in Gansu, the water control projects stood like a row of dominoes.[26]

After the lower-lying lands had been irrigated, the people of Gansu turned to building even larger pumping facilities to water the drought-stricken high plains. In 1969 provincial authorities spent a large amount on the country's first high-lift, multilevel pumping station in Jingtai and Gulang counties. Then it went on to spend billions on pumping Yellow River water to nurture "another Gansu."[27]

While pulling water from the Yellow River certainly helped raise the standard of living of Gansu's people, it had a negative effects on the river itself, as the water taken from the main channel no longer reached the middle and lower portions. Downstream there were howls of outrage about the low water efficiency and the astronomically high costs.

In 1994 Gansu used 4 billion cubic meters of river water; in 1997 it was 4.2 billion cubic meters. That greatly exceeded the limit of 3.04 billion cubic meters that had been set by the central government.

In return for this life-sustaining substance, Gansu gave the Yellow River an immense amount of fine yellow dust from the loess sediment that covered the landscape. Gansu is estimated to discharge 518 million tons of sediment into the river annually, or approximately a third of the sediment carried in the main channel.

The people of Gansu are perhaps even more frightened of soil erosion than people living along the lower reaches because they have lost such large amounts of fertile land to this inexorable process.[28] Sediment has accumulated to such a degree that the major hydroelectric power stations in Gansu have had serious reductions in generating capacity—as much as 25 percent at Liujia and 84 percent at Yanguo Gorge. Power stations that required an immense amount of investment to build are now virtually worthless.

A person would be amazed to see just how carefully the local people conserve water—half a basin with which to wash their face, then wash the dishes, then slop the pigs, and then (if anything is left over) water the few trees and shrubs struggling for existence.

It's impossible not to feel sorry for these people, but they're the same people who would not hesitate to flood a patch of parched ground in the most profligate way the minute they had access to Yellow River water from reservoirs or pumps. Studies have shown that each peasant family was responsible on average for 15 tons of lost soil.

Gansu proved that old water management methods were a total failure. The solution, of course, seemed rather simple: stop cutting the trees, stop turning barren land into irrigated land, and stop the excessive use of water resources. But if people did not resort to these methods, how could they possibly make a living? Any complete solution would take a great deal of courage—but also an immense amount of brainpower.

THE YELLOW RIVER: ENRICHER OF NINGXIA

Water flows through irrigation ditches down to the land and the water-driven mills help them husk the rice. [Ningxia p]eople don't have to do much to have a good harvest.
　　　　—*History of the Eastern Han* (A.D. 90), Ban Gu

THE YELLOW RIVER RUNS FROM QINGHAI'S Longyang Gorge to Ningxia's Qingtong Gorge for 916 kilometers through nineteen valleys toward the Yinchuan Plains. The plains calm the waters of the Yellow River on their more than 21,000 square kilometers. There's a gradient differential from south to north of almost 200 meters. In many places the farmers can simply open the river's levees and let the water run down into irrigation ditches and canals and onto their land.

Drawing water off the calmest part of the Yellow River has long been a practice in China. It started in Ningxia and was constantly expanded.[29]

For thousands of years, the people of the Yellow River valley were grateful for the wealth the great waterway brought them. When the Yellow River began to dry up, all eyes turned toward Ningxia to look for the culprits. Who was abusing the river water with wasteful flood irrigation methods, instead of the more frugal drip method?

A Yellow River water distribution formula was developed in 1987, and Ningxia was to receive 4 billion cubic meters annually, or about a ninth of the total river flow. Ningxia was China's smallest province, with less than 3 percent of the total number of people living along the entire Yellow River valley, so the share was quite generous. However, data from the Qingtong Gorge Dam, the largest water supply project on the Ningxia section of the river, showed that in fact Ningxia had diverted about 6 billion cubic meters of water annually (more likely 8 billion cubic meters).

When people living along the parched lower reaches heard this they were furious. But the Ningxia people had an explanation. They didn't deny that they had used 8 billion cubic meters but said that after irrigation needs were provided for, about half of it, or 4 billion cubic meters, was returned to the main channel of the Yellow River.

They also had their own protest against the provinces along the lower reaches, such as Henan and Shandong, saying they should be held responsible for the drying up because they had been returning almost no water to the main channel. In fact, because the river in those parts rose above the surrounding landscape, there was no feasible way to put unused irrigation water back into a suspended river.

The "blame game" went on and on. In reply to the claims of the Ningxia people, those living along the lower reaches said that when the 8 billion cubic meters of water was drawn from the Yellow River it was relatively clear, but when the residue was returned, it contained chemicals, fertilizers, sewage, and industrial waste. They pointed out that Ningxia had increased its irrigated land more than threefold over the previous fifty years and urged those people to stop the reclamation craze.

The people of Ningxia were unmoved and argued in return that even more irrigated land downstream was reclaimed land and that the middle and lower reaches were using seven times more water for irrigation than they had been fifty years before.[30] As long as the figures remained that high, the Ningxia people pointed out, it wouldn't matter how much water they got from upstream because it would all be consumed anyway.

Confronted with this difficult fact, people along the lower reaches argued more fervently, saying that they used the water much more efficiently than the people they were feuding with, and blamed Ningxia for wasting vast amounts of water in its broad flood-irrigation methods.

Ningxia's reply was that they were trying every method they could find to save water and, according to their calculations, the methods they had applied over the previous several years had saved upward of 300 million cubic meters. They had to admit, however, that that was but a drop in the ocean.

There was finger pointing back and forth, working people at both ends of the trail of blame into a frenzied panic and impelling them to move ahead with their own water projects in a crazed way. Almost every province and region has poured billions of yuan into major projects that would give it an edge in getting water from the Yellow River. Ningxia was one of many.

Downstream, where the Yellow River heads northward out of Ningxia into Inner Mongolia, there is the key water diversion project of Sanshenggong. It was built in 1959, during the Great Leap Forward, and turned an arid pastureland into one of the most heavily irrigated agricultural areas after Ningxia. It has a huge canal intended to divert water from the Yellow River that is more than a hundred meters wide and which much of the time actually carries more than the main stream of the Yellow River itself.

More than 5 billion cubic meters of Yellow River water are siphoned off annually in and around this Great North Bend area, irrigating 570,000 hectares of land. But, thanks to a lack of repairs and poor maintenance on the diversion project, only 60 percent of that transferred water is used, meaning that every year, 2 billion cubic meters of precious water is lost through seepage.

In 1992, Inner Mongolia siphoned off as much as 7 billion cubic meters of water. Only a very small amount of rainfall hits the grasslands of that region in any given year (less than 300 millimeters), and with an evaporation rate of a whopping 2,000 millimeters, much of the Yellow River's water ends up being scattered across grass and yellow sand. Only a billion cubic meters makes its way back to the river.

Altogether, the flood irrigation methods in Ningxia and Inner Mongolia cover more than a million hectares of ground. Annual water usage is 10 billion cubic meters. What begins as a powerful river at the edge of Qinghai province is little more than a stream by the time it traverses Inner Mongolia.

So the question naturally emerges: How can one expect the Yellow River, under these conditions, to last for 3,000 more kilometers and make it to the sea? And what, if anything, can be done? Unfortunately, asking the people of Ningxia or Inner Mongolia to give up their water rights is no more feasible than telling the people of Gansu to stop relying on the Yellow River to make their living. This is the crux of the matter.

There's another sad fact embedded in this tale of woe: In spite of the profligate amounts of water drawn from the river, overall the environment of Ningxia and Inner Mongolia has shown little improvement. Around the Great North Bend area, as in many other parts of northern China, the green cover is in retreat and is rapidly being replaced by the yellow sands of the encroaching desert.[31]

The Yellow River flows for a mere 390 kilometers through Ningxia, and after the province draws off water for its own use, it gives nothing in return but pollution and 50 million tons of mud and sand. The Yellow River brings wealth and prosperity and gets significantly degraded in return. Isn't it clear that our water management is nothing but a failure?

THE YELLOW RIVER: PROVIDER OF DEVALUED WATER

If you don't grab it, someone else will. If a person misses out on anything free, he looks like a fool.
        —Chinese saying

CHINA RANKS 122ND OF ALL THE COUNTRIES IN THE WORLD in per capita water usage, at 2,300 cubic meters per year, or a fourth of the world average. Along the entire course of the Yellow River, per capita usage is only a fourth of that national average.

The Yellow River has only 2 percent of China's water resources but has the burden of providing upward of 15 percent of its irrigation water and 12 percent of the water the country consumes. In all, 53 percent of the entire river's water goes to human consumption, a figure that is not paralleled in any of the world's other major rivers.

The Yellow River water is invaluable, but do the Chinese treat it with the respect it deserves? The fact is that for thousands of years, humans in China have wreaked havoc on the river, with just about everyone along its banks exploiting it and abusing its water. These days, thanks to modern engineering advances, even people far from its banks have found ways to take advantage of its resources.[32]

Projects have been proposed and authorized with little regard for the overall interests of the nation and certainly not those of the river itself. But the river is simply in no position to take on any additional burden or allow a drain on its dwindling resources.

Increasing the amount of irrigated land along the upper reaches will only cause suffering along the middle and lower reaches from the increased drying up. If more of its precious water is used along the upper reaches to make desert land fertile, the middle and lower reaches will suffer from increased desertification themselves.

There are more than 160 water projects already built along the lower reaches that can divert a total of 4,100 cubic meters per second. But average water flow in that part of the river valley is only 1,839 cubic meters per second.[33]

The question arises: If we are merely tearing down the east wall to build up the west wall, why is it that more dams and water diversion projects are being planned? The reason is simple: Local people all up and down the valley are suffering and see the projects as the only means of survival. They have little room for thought of the impact on people farther down the valley, let alone on the rest of the country.

Even though Yellow River water can be said to be invaluable, there is a price on it, and it is incredibly low—almost nothing, in fact. It runs about 0.003 to 0.004 yuan per ton. A thousand tons of Yellow River water actually cost less than a single bottle of mineral water. There has been a tremendous undervaluing of this now precious water.

After fifty years of construction of these major water control projects, the Yellow River has just about reached its limit. From the early 1980s to 1993 the total amount of water diverted from the river annually was 27.6 billion cubic meters, four times the amount drawn off prior to 1949.

Look at any map and the Yellow River still appears like a large squiggly line. In reality it has been reduced to a series of narrow streams combined with a few reservoirs. In spite of the fact that the country's constitution says that rivers are state property, there is little in the way of specific regulations or guidelines for the use of their waters.

Although the economic reforms inaugurated by Deng Xiaoping in 1978 have been in place for more than two decades and have led to the introduction of a market economy, as far as the allocation and utilization of the Yellow River's water goes, the country might as well be back in the age of the "big iron rice bowl," where everybody dips into the communal socialist pot because little consideration is given to the real cost of water.

In Shandong and Henan provinces, a study that was done in 1995 revealed that the prevailing price of river water for irrigation was a mere 0.002 to 0.003 yuan per cubic meter, while upstream in Ningxia it was 0.005 yuan per cubic meter. But in Ningxia a kilogram of rice took 2 cubic meters of water to grow, and in Inner Mongolia the same amount of grain took 6 cubic meters, while the national average was about 1 cubic meter.

In the same way the agricultural reforms that also began in 1978 proved that eating from the "big iron rice bowl" was not feasible, it is time to realize that drinking from the "big iron water bowl" is impossible. And yet the fact remains that along the middle reaches of the river, water prices remain incredibly low. This situation cannot continue.

The Ningxia irrigation district provides an example. Since 1989 the price of water has been adjusted upward two times, to 0.006 yuan per cubic meter for flood irrigation and to 0.05 yuan per cubic meter for pumped irrigation water. In 1997 total revenues from flood irrigation water use were just under 60 million yuan, while the actual cost of the water was almost twice that. When we consider that overall efficiency in the use of Yellow

River water for irrigation is only 40 percent, then, for every 30 billion tons of water diverted, 18 billion tons are lost to seepage or elsewhere.

No matter what the solution to this enormously wasteful situation, most experts agree that if efficiency could be increased to 60 percent, 6 billion cubic meters of water could be recovered each year. And if efficiency were increased to 90 percent, the amount of water saved would be 15 billion cubic meters—almost equal to a second Yellow River. It seems logical that more should be done to get people to see the true value of Yellow River water.

A CALL FROM AFAR

Beautiful this place is, with the big lake and the meandering clear river.
—Helian Bobor, a khan of the Huns, who built a capital on the Ordos grassland in A.D. 407

CROSS THE IRON BRIDGE over the Yellow River at the city of Baotou in Inner Mongolia, and the road immediately turns south toward the vast Ordos, an arid plateau surrounded by the river on three sides.

In the water's surface, one can catch the reflection of puffy white clouds and an azure sky. On both banks of the river thousands of sunflowers stand golden in the sun. Passing through this area, a person might easily imagine that here, at least, the water is pure.

But just ask the locals. They claim that it tastes funny—funny enough that anyone with a high position or a good income relies on bottles of mineral or purified water for washing vegetables and cooking.

If one goes a bit farther along the Yellow River, it becomes clear that one side of the road is relatively smooth and the other quite bumpy. That damage is the result of the constant use of the roads by large, heavily overloaded coal trucks.

This is one of China's major mining regions, an area that provides a very high grade of coal. There are new mines at Dongsheng and Zhunge'er, built by the state at a cost of 20 billion yuan—great for the coal business, but the impact on the local ecology has been devastating. The major recipients of that treatment have been the trees and the vegetation. Their destruction has, as in other parts of the Yellow River valley, left very serious soil erosion problems in its wake. According to the Yellow River Conservation Commission, the outlook for soil regeneration in this area — known, not surprisingly, as the Black Triangle — is not especially rosy.

From 1990 to 1996 new coal mines were opened, reducing the rate of runoff into an area of catch basins by 20 percent, accompanied by a 30 percent increase in the amount of sediment compared to the years before 1990. The coal in the Black Triangle is very easy to mine because it lies close to the surface. But the open-pit mines have left an increasing number of major sinkholes in the surrounding landscape, and the insulating topsoil has been so seriously eroded that there has been a major drop in the water table and a corresponding loss of plant and tree cover over large pieces of ground.

One investigation showed that a single mine had dumped 3.3 million tons of slag on the area annually. Some news reports have revealed that the Black Triangle is responsible for some 396 million tons of sediment entering the Yellow River annually.

The problem of soil erosion in the Ordos Plateau has been exacerbated over the years by contact with a type of rock known in Chinese as *pisha*. When it is covered by the yellow soil that is prevalent in the region, pisha is no different from any other type of rock. But when that soil cover erodes away, exposing the rock to the wind and water, its loose structure gives way. When the Ordos Plateau is pelted with heavy rain during the summer, the rocks break up and join the runoff into the Yellow River area, making their way eventually to the river itself.

The amount of sediment containing pisha that enters the river is astonishing.[34] To make matters worse, most of the sediment consists of coarse sand that sinks to the bottom of the river, where it gradually helps raise the Yellow River.[35]

This now miserable place was a beautiful pastureland when the first emperor of China took it from the Huns. The historian Sima Qian wrote that the Ordos had some of China's best pastureland. The ambitious Emperor Wudi, who reigned during the time of Sima Qian, moved nearly a million peasants to the area to turn the grassland into farmland to increase the country's grain supplies. Centuries later, during the powerful Tang Dynasty, troops were stationed there to open up more land.

But after Chinese farmers fell under attack by barbarians and fled the place, the Ordos returned to grassland. Eight hundred years ago Ghengis Khan was so attracted to the beautiful lakes and grass that he set aside a piece of grassland for his burial plot.

After the Mongol Empire was toppled, China once again turned to reclaiming the vast plain. But the degradation that followed that expansion pales in comparison with the two large-scale land reclamation campaigns of the 1950s and 1960s. These fantastic projects left in their wake two small deserts that expanded and finally embraced each other in the 1970s.

Today the great khan's tomb is still there, cared for by descendants of his Mongol guards. But the green grass and beautiful lakes are gone. They were replaced by an endless stretch of sand.

So by the time the great river exits Inner Mongolia, it is carrying an immense load of silt, which over the centuries has brought despair to people living along the middle and lower reaches.

STUNTED GREENERY ACROSS SHAANXI

No one around here is afraid of falling into the muddy river.
—Local saying in the Wuding River valley

AFTER IT CROSSES THE RELATIVELY FLAT AREA of the Great North Bend and the Ordos Plateau, the Yellow River leaves Inner Mongolia, flowing in an easterly direction, but is blocked by Luliang Mountain. So it turns nearly 90 degrees south to the Chinshen Valley, between Shaanxi and Shanxi provinces, an area that has been carved by centuries of wind blown dirt and eroded loess and where hundreds of tributaries feed into it along a 700-kilometer stretch.

There is more rainfall here, and the Yellow River recovers some of what it lost getting here, with the flow going from 24.8 billion cubic meters to 49.7 billion cubic meters.[36]

The middle reaches are essential to the river's survival. It cannot expect any further help downstream, where it becomes a suspended waterway. But unfortunately, even here the environmental damage on both sides of the valley means less water and more sand for the river.

We first take a look at conditions on the left bank in Shaanxi province. In the northern part of this province the desert that follows the southern edge of the Ordos Plateau meets the eroded parts of the Loess Plateau with their deep ravines and gullies. Although there are few tributaries in this region, the amount of sediment running into the main stream of the river, whose gradient in the loess region is greatly increased, is quite astonishing.

The Kuye River, in the northernmost sector, is a fairly small watercourse but dumps as much as 130 million tons into the Yellow River every year. To its south is the equally silt-laden Tuwei River, with even less water than the Kuye. But once it is flushed with the periodic summer floodwaters, the Tuwei turns into a raging torrent of mud, all of which ends up in the Yellow River.

Then there is the Wuding, just south of the Kuye and Tuwei. It is also a relatively small river, no more than 490 kilometers long. For years it has had an average of 138 kilograms of sediment per cubic meter. Sometimes it reaches 1,520 kilograms per cubic meter, a load that might rival that of any of the world's rivers.

According to Yellow River Conservation Commission employees, there was a huge flood some years ago, after which the amount of sediment in the river was so large that young peasant boys could be seen bobbing along in it without the slightest fear of sinking. Hence the comment "No one around here is afraid of falling into the muddy river."

Northern Shaanxi was once a richly lush area, but its ecosystem began to be seriously compromised during reclamation campaigns of the Tang Dynasty. The area was famous as a site of battles between warring tribes, but Yulin is now the front line of a more difficult battle that has been waged since 1949, one with little to brag about and no victories. That is the battle against the encroaching desert.

By 1998, trees had been planted on more than a million hectares of land to try to halt the advancing desert, but that is hardly significant since overall forest coverage in the region is still very low, hovering around 4 or 5 percent.

And while the amount of sediment in the Kuye, Tuwei, and Wuding rivers has been dramatically reduced in recent years, there is, in fact, little call for celebration. The fact is that most of the sediment reduction was a consequence of the serious drop in water flow in these three rather minor streams. And that came about largely because of the increasing frequency of droughts that are themselves a reflection of the climatic changes in the region.

As we go south from the Wuding River, we say goodbye to the grasslands of the Ordos Plateau and enter the loess area of northern Shaanxi. This is not only the central axis of the Loess Plateau but also the main area of soil erosion. The Yan River, which flows through the region, is a tributary of the Yellow that has gained fame as a part of the revolutionary history of the Communist Party. It runs through a sort of holy land, Yan'an, the party's capital and revolutionary base prior to and during World War II.

Pagoda Mountain, a symbol of the Chinese revolution, can still be seen here. But the Yan River, another revolutionary symbol, has not been so fortunate. The river has been reduced from the torrent it was sixty years ago to a thin muddy stream after years of serious soil erosion.[37]

Further south on the trek through Shaanxi, we enter the valley of the Wei River, 818 kilometers long. The Wei, when combined with the Jing and Luo rivers—two of its own tributaries—drains a total area of 170,000 square kilometers and has a total capacity of 920 million cubic meters, making it the largest branch of the Yellow River.

The entire area drained by both the Luo and Jing rivers has severe soil erosion problems.[38] The Wei has become the largest single repository in the entire basin. But for thousands of years it provided nourishment for the historically important city of Xi'an, which was the Chinese capital during the glorious Qin, Han, and Tang dynasties. Xi'an gained even greater fame over the past two decades as the home of the terra-cotta warriors put there by the Qin Dynasty's Yellow Emperor.

But that was then. As the river gradually lost its water Xi'an turned increasingly to subterranean supplies to meet its water needs. The city now has more than a thousand wells, but its thirst for water still cannot be met. Its daily shortfall is estimated at 500,000 cubic meters.

As the water table dropped because of that increased demand, cones of depression (funneling) appeared. Since the 1960s, the ground around Xi'an has been sinking at an average of

34 millimeters a year. The Big Goose Pagoda has sunk 1.3 meters and now tilts 1 meter to the northwest, giving it new prominence as China's own great "leaning tower."

A more serious threat, though, comes from the dozen or so major fissures and sinkholes that have popped up all over the city, causing houses and buildings to collapse and natural gas pipelines to rupture, resulting in explosions, all of which now put the city's residents at considerable risk. So the water crisis is tied to Xi'an's ultimate survival in every possible way.

In its search for the ultimate solution to Xi'an's chronic water problems, the municipal government resorted to extreme measures, boring through the Qin mountains to divert water from the Hei River. Problems began cropping up almost immediately. But some had been evident as early as the 1950s.

The water volume of the Hei had already dropped 32.6 percent as a result of deforestation in the Qin mountains. When the water diversion project was completed in the 1970s, the Hei began to experience a severe reduction in water flow. That left Xi'an with no alternative but to seek additional water from the Shitou River. That round of diversion cost 3 billion yuan and, hardly surprisingly, created problems for the Shitou, which, like the Hei, saw its total volume plummet by 23.6 percent from the 1950s level.

This then was yet another vicious cycle that took its toll on the region's water resources. Water diversion projects were built helter-skelter with little or no regard for the cost, while at the same time the continuing destruction of the forests that offered the greatest protection of the water resources went unobserved.[39]

Soil erosion in Shaanxi province currently affects some 137,500 square kilometers of land, or nearly half the total amount of eroded land in the entire country. It also produces 920 million tons of sediment annually on average, or a fifth of the total amount of soil erosion in the entire country. This situation has been made worse by the droughts that have plagued Shaanxi over the past several decades.[40]

Shaanxi's vice governor, Wang Shousen, was left to lament: "Nothing has hampered the economic and social development of Shaanxi more than the water shortage."

In view of all this destruction and the traumatic struggle for survival in Shaanxi, it's not surprising that the local government lobbied the central government for help in building large dams that would give it more water from the Yellow River. Obviously, it was pointless to try to enlist the beleaguered citizens of Shaanxi in the fight against the Yellow River's water crisis.

SHANXI'S BEAUTIFUL SCENERY

Floating on the vast Fen River, my mighty boat pushes a white wave.
—Emperor Wudi (140–87 B.C.), Western Han Dynasty

WE LEAVE SHAANXI NOW, going across the Chinshen Valley, from the left bank to the right, to consider the conditions from the Shanxi side.

Shanxi is a sort of landlocked unit all its own.[41] The 700-kilometer-long Fen River is the province's longest, and its valley was the actual cradle of Chinese civilization, with verdant forests, fertile soil, and abundant rainfall. The path of its ecological destruction is also the most prolonged. It dates back almost 4,000 years.[42]

The area is now largely void of trees and lesser vegetation, and the Fen's annual flow has been reduced. Much of its water ends up in the Fen River Reservoir, which has a capacity of 710 million cubic meters. But, as has already been explained, this one has lost much of that capacity to sand and sediment.[43]

Not far below the reservoir and just downstream from the provincial capital, Taiyuan, the Fen has no water. Most of what can be seen flowing in it is sewage.

The Fen River Valley is Shanxi's most important industrial and agricultural area and is perhaps best seen from a train, which affords an astonishing view of a completely broken-down landscape. The hillsides are all barren and etched by an endless array of cracks and gullies. The towns and villages are uniformly gray and punctuated by rows of tall chimneys belching out black smoke that wafts down each small valley and ends up in the larger one.

Each of the streams that join the Fen, coming from towns and villages up in the hills, has its own unique color, depending on what sort of malevolent witches' brew the local township factories are releasing into it. The streams run black, orange, various green hues, rusty reddish brown, or blood red, making the valley look like a vast artist's palette.

The Fen is the mixing bowl for all the colorful tributaries, and it creeps along clinging tenaciously like a small ribbon to the parched riverbed. Inside the passing train coaches, a voice comes over the sound system in a mesmerizing tone, telling visitors of Shanxi's wonders, of its beautiful scenery, and of a dreamy place with fertile land and clear rivers.

Statistics show quite clearly that the once glorious Fen River dried up because of the growing numbers of people and industrial and agricultural use. In the past five decades, total water use in the province has gone from 500 million tons to more than 7 billion tons.

The local people never let up, and the ecodestruction was unrelenting. First they farmed the hillsides; then came the coal mining and after that the other industries. Ore was processed in the area because it was convenient to build the plants close to the mineral source. The runoff from water sources in the province dropped from 600 million cubic meters in the 1980s to 500 million in the 1990s.

The Fen's water sources began to disappear and the large number of springs that used to dot the valley began to dry up. Still, the need for water meant digging further down to draw off the underground water supplies. According to some studies, each year the people of the area drew off nearly 2 billion cubic meters of water from the aquifer in the Fen River Valley to provide 86 percent of the year's water supply.[44]

What about those many small rivers in Shanxi that slowly dried up over the past several decades? Most are now nothing but open sewage drains, while the waters of the Fen, the Dahei, the Laishui, and the Mang have so many contaminants that they have been classed as some of the most polluted bodies of water in all of China.[45]

And the pollution in the Yellow River? The main source of it now is the Fen, although it must be noted that Shanxi is by no means the only territory using the Yellow River as an open sewer.

There was the horrific case that occurred in January 1999 near where the Xiaolangdi Reservoir construction work was going on. A layer of foam half a meter thick spread across the river channel and down for several kilometers, polluting the water so badly that purification plants in many large cities in Henan and Shandong provinces had to shut down, cutting off the water supplies to several million people.

This egregious spill was taken extremely seriously by provincial and national authorities, who were determined to ferret out the culprits. That search, unfortunately, came to naught because finding a single source of the contaminants proved to be difficult if not impossible. The reason was that almost every single stream along the Yellow River—from Qinghai to Gansu, to Ningxia, to Inner Mongolia, to Shaanxi, Shanxi, and Henan—was a source of pollutants that eventually made their way into the main stream. But the investigation continued and after a lengthy period the source of the majority of it was traced to the Fen River.[46]

At present, the section of the Yellow River that is downstream from Shanxi has the worst possible water quality rating there is (4), and even above that level, which means the water is not suitable for any human use. The Yellow River is estimated to carry 5 billion tons of untreated wastewater. During the dry season, when freshwater levels are at their lowest, the ratio between wastewater and freshwater is as high as 1:10.

Strangely, the Yellow River has long been thought of as being the only hope for the Fen. After getting a loan from the World Bank, the province spent 14 billion yuan on the Wanjiazhai water diversion project, whose purpose was to take water from the Yellow River and put it into the Fen. Of course, the Fen is a tributary of the Yellow and one of that river's major sources of replenishment, so taking water from the Yellow to put it in the Fen would obviously be a matter of tearing down the east wall to build up the west wall and no doubt would cause the Yellow River to dry up more. It is hardly surprising, then, that this proposal caused more howls of protest from people downstream in Henan and Shandong.

At the lower end of the Chinshen Valley is the Hukou Waterfall, one of the most exciting places in the entire Yellow River valley. Just above the falls, the river goes from 400 meters wide to only 100 meters and the waterfall plunges furiously down

30 meters over limestone steps. The roar of the falls is so loud that peasants claim it can be heard 15 kilometers away.

In recent years, some daredevil came up with the idea of jumping the falls. The logical vehicle for this new fashion was a motorcycle or a really hot car. In 1997 a Hong Kong stuntman made it across in a racecar. Two years later a local boy, Zhu Chaohui, got his face in the papers by pulling the same trick on a motorcycle. Next door, in Shaanxi, when they heard about this, some folks decided to join in the merriment and said they too could leapfrog the falls — this time on a dozen bicycles.

Some people were excited about the idea, some said it was downright crazy, and some were skeptical. But one wise man pointed out that the trick was indeed possible for the simple reason that the water flow at the falls was so incredibly reduced, especially after Wanjiazhai, that the place they had to jump was just about suitable for a bicycle.

And so it goes. The Yellow River's water level is being constantly drawn down by various diversion projects on the upper and middle reaches, and the shrinkage will continue. So someday we will be able to jump the falls without a motor of any kind. And just how will we feel? Proud? Or ashamed? This is, after all, the "mother river."

FIFTY YEARS WITHOUT A MAJOR BREACH

When floodwaters rise an inch, we will raise our embankment by ten.
    —Official slogan during the record flooding on the Yellow River in 1958

IF ONE TRAVERSES ALL NINE OF THE PROVINCES touched by the Yellow River, one can perhaps only come away dazed by the situation and the knowledge that its particular problems have not been created by one particular province, particular historical period, or even particular project.

Instead it is necessary to think cumulatively of the destructive processes along the upper, middle, and lower reaches that have gone on for thousands of years, and which more recently have come from untamed growth while the water resources remain limited.

Many people have commented on how the single-minded devotion to huge dams and other projects led to chaos and the drying up of the Yellow River, while most others have stuck to the benign view that they are efficient in controlling the floods of an insanely violent river.

From 602 B.C. to A.D. 1938, there were 1,590 major collapses of Yellow River dikes, usually once every two or three years. Then, every hundred years or so, the river would change its entire course, wreaking the most awful havoc with remarkably destructive flooding.[47]

But then these inexorable and unpitying cycles seemed to come to an abrupt end after the Communist Party turned its hand to river management in 1946. And it so happened that though each autumn from 1946 to the present brought a renewed threat of major floods, none has breached the river's defenses, leading people to speak of "fifty years without a major breach."

To see to it that the river would no longer go over its dikes, the new government spent an inconceivable amount of money and manpower on flood control projects. This narrative has centered on the 3,000 dams up and down the river, but the embankment is a story in itself, a whopper.

Since 1949 and the establishment of the People's Republic of China, there have been three major projects to improve the levees and embankments along the Yellow River. These projects were colossal and colossally expensive. They involved the equivalent of 500 million workdays and 1.4 billion cubic meters of reinforced concrete. That is enough to build 13 Great Walls.

These massive engineering projects did help to bring the violent river under control. But did they really get at the root of the problem? That is the question.

In 1998, the Geology Bureau of the Chinese Academy of Sciences joined local specialists in organizing a group to study the Yellow River in its entirety. In short, their findings astonished just about everyone. Not only were the embankments and dikes of the Yellow River in danger of a collapse worse than any since 1949, but also there was the possibility that the mighty river might change its course, with catastrophic consequences.

The reasons for this were obvious: After years of decreased water flow in the lower reaches, together with the prolonged dry periods, the pattern of sediment accumulation had changed radically, with most silt having accumulated in the main riverbed during the preceding decades, as opposed to the wider river basin and estuary.[48]

To make matters worse, the river was gradually being encircled with new landfill. Here it is helpful to recall that in 1958, during the frenzied behavior of the Great Leap Forward (1958–60), people living on both sides of the river were ordered to march into the river valley and fill in the spots where there were depressions, and eventually settle and farm the area.

The result of this irrational behavior was that the river lost many of its natural catch basins and no longer had places inside the embankments where all the sedimentation could be deposited. Gradually the bed of the main channel rose higher than the sides, which were several meters higher than the surrounding plains. This gave the river its unique "two-tiered suspended river" look. Two million people living along the river in the valley are today very vulnerable.

High embankments and large dams helped turn smaller floods into big disasters.[49] If a flood similar to that of 1958 were to occur now, the water level of the river would be 2 to 4 meters higher than it was then.[50]

In the lower reaches, the river's course is wider at the beginning and narrower at the end. There are dangerous conditions in both the widest and narrowest sections.[51]

The easiest way to deal with this problem would be to divert floodwaters into Dongping Reservoir and the drainage basin.[52] But one must remember that Dongping Reservoir was designed and built during the Great Leap Forward and can be considered neither scientifically sound nor of particularly good quality. That kind of thinking can only cause worry and concern for anyone who understands the current Yellow River situation.

There are those people who believe that the root of the problem is the fact that thirty-six separate water control projects planned for the main stream in 1955 have still not been completed. Once all are finished, this line of reasoning goes, the capacity along that entire stretch of the Yellow River will come to more than 100 billion cubic meters. At that point, controlling the Yellow River will no longer be a problem.[53]

A different point of view is held by those who argue that there are already a large number of water control and diversion projects on the Yellow River and that if even more dams are added, the river will no doubt become nothing more than a thin stream and an ecological mess. More reservoirs, these people say, will simply move the Yellow River farther and farther away from the sea and make an already bad situation worse, with sediment and silt accumulating in the lower reaches and the river drying up.

Some historians have pointed out that fifty years without a major breach is not especially unique. There are records of the Yellow River actually flowing rather peacefully for as long as 800 years, from the Eastern Han Dynasty (A.D. 25–223) to the Tang. They argue that it had little to do with the embankments because the Chinese in fact had lost control of much of the river basin to barbarians.

The explanation of the miracle, then, is that the Chinese abandoned all their farmland in the Ordos and on much of the Loess Plateau and the natural vegetation gradually came back to a vast area, halting the unnatural soil erosion and the silting up of the lower reaches. When the forces of nature followed their course without humans intervening and attempting to harness the flow, the river stuck to its original contours.

The question then is: At this juncture in history, practically at the dawn of a new millennium, why cannot human beings give up their ruthless ambition of harnessing and controlling nature and choose instead to live in harmony with it?

DEATH OR SURVIVAL?

Only when the barren hills are covered by lush forests will we see the Yellow River turn clear.
——From a petition drawn up by 163 prominent Chinese scientists in 1998

THE ONLY QUESTION LEFT, AND THE WHOLE POINT OF THIS NARRATIVE, is this: What can we now do to rescue this great waterway, the "mother river," after it has been proven that huge dams and large embankments will only kill it? More and more, people have been looking for more environmentally friendly solutions, one of which is the reforestation of the Loess Plateau and which all of a sudden seems to be a priority.

While this is an encouraging development, it cannot be taken for granted. In fact, various slogans about "greening the Loess" have been mouthed for decades, so it might be better to try to figure out exactly how or where or why the attempts went wrong.

First, there is the lack of funding, a perennial problem. It has been with us for years. The total amount spent on soil conservation — 2.3 billion yuan since 1949 — accounts for less than 4 percent of all state spending on "harnessing the Yellow River."

A further problem is the fact that the spending was never really applied to viable programs. That paltry amount, when spread over the vast Yellow River basin, might as well have been scattered from an airplane and allowed to fall where it might.[54]

The soil erosion is out of control and has contributed to the silting up of large reservoirs. The accumulated sediment in existing Yellow River reservoirs amounts to 13.4 billion cubic meters, or enough silt, sand, and gravel to fill 134 large reservoirs.

The most encouraging new development is the government's determination to spend billions of yuan to stop the farming of hillsides. In 1999 it began ordering farmers living near the headwaters and on the Loess Plateau to put a stop to it. The government has provided compensation in the form of cash, grain, and saplings.

The reason for this is that experts, in studying the middle reaches, have pointed out that no attempt to grow wheat or corn in the area would lead to much benefit when the environmental cost is factored in. If what water there is were to be allowed to run on down to the North China Plain, farmers there could grow several times more grain and cotton and factories there could have greater output with the same amount of water.

The government compensation provides an escape mechanism for farmers stuck in a vicious cycle. What is now needed is a mechanism whereby the provinces downstream that benefit can compensate those on the upper reaches to provide reforestation capital.

In the case of the drying up, nothing can be done unless it is determined precisely how much water can be diverted from the river as it now stands. This, obviously, is easier said than done, especially during a period of economic and governmental reforms, some of them radical, where the old planned system can no longer be relied upon for coordinated policy or action. Here, market forces need to be called upon.

Simply put, this means that the price of Yellow River water must be increased dramatically to reflect its true market and resource value. In a market economy, price is the "invisible hand" and is believed by many, though by all means not all, to be the fairest way to allocate scarce resources. If a person wants to have the benefits of any resource, then the appropriate amount must be paid. This is the only workable solution in confronting the dire situation on the upper, middle, and lower reaches.

The current water price is so low that farmers, factories, and urbanites could afford a rather substantial price hike. In the past, when the price of irrigation water doubled in several provinces, farmers coped with the change by adopting water-saving methods. In Ningxia, where the water price went from 0.006 yuan to 0.012 yuan in the year 2000, more than 600 million cubic meters of water was saved that year alone.

Increasing the price of water will also help reduce the excessive and wasteful use of water by industry and just might have the additional benefit of reducing water pollution from industrial sources. Factories up and down the river currently recycle only 30 percent of the water they use.

Greater controls are also needed on the almost unlimited demand for water by 20 million urbanites, who need to be made more aware of the fact that this scarce resource cannot be used so profligately. Australia, Canada, Israel, and the United States have demonstrated that for every 10 percent hike in the price of water, residents cut demand by 3 to 5 percent.

Higher water prices would make sewage treatment plants economically viable, and market forces might provide more encouragement for the use of nonpotable water, including seawater, for things such as toilets.

If price reforms take effect, the government will need to establish a regulatory body to ensure that Yellow River water resource controls are applied in a unified fashion. Water will always be in demand, even as its price increases, so applying water rights controls is inevitable.

The distribution of Yellow River water to the nine provinces and regions through which it flows should accord more with the principle that those that do the most for the river have greater rights. This includes those provinces and regions near the source that deserve more.

China is now engaged in its campaign to develop the western part of the country, and provinces such as Qinghai, Gansu, Ningxia, Inner Mongolia, and Shaanxi are trying hard to open up new farmland and build more factories. This will do more ecological damage to those fragile regions and produce more industrial waste, which will flow down to the coastal regions.

If Qinghai province is given the water rights it deserves, it may choose to profit by selling those rights instead of wasting water on its parched land or on inefficient industrial projects. Both its ecology and the local economy would benefit, and that, in turn, would benefit the Yellow River. When farmers choose to stop their destructive hillside farming practices along the upper and middle reaches and decide to sell their water rights to people downstream, we may one day see the return of vegetation to barren hills and even see the Yellow River run clear.

For thousands of years the Chinese have prospered along the banks of the Yellow River, but at the expense of the vitality of the waterway that was supporting them. Now it is up to us to decide whether we will follow in the footsteps of our forefathers, living with that paradox and making the same mistakes.

The survival or death of a once mighty river is the option. If the Yellow River dries up in front of our very eyes, the shame and blame will forever be ours. But if we can bring this mighty waterway back to life, it will be our greatest glory.

CHAPTER TWO

THE YANGTZE RIVER

"GREAT RIVER FLOWING EASTWARD"

It was alongside a river where Confucius said: Thus do things flow away, never ceasing day and night.
—Analects (436 B.C.–402 B.C.), Confucius

PERHAPS NO SIGHT COULD BE MORE SOOTHING, or more interesting, to the eyes of someone who is jaded after seeing so many parched riverbeds in various parts of China than the wide, swift, and turbulent waters of the mighty Yangtze.

This powerful river begins in glaciers 5,500 meters up in the Gelandandong Mountains on the Qinghai-Tibetan Plateau and flows across central China, absorbing a vast number of branches and streams, and finally spills into the East China Sea just north of Shanghai. It is the pulse of the nation.

A fifth of China's entire land mass is drained by the river. It flows 6,300 kilometers through eleven provinces, while its more than 700 tributaries embrace eight other provinces. The river has a 1.85-million-square-kilometer catch basin, which is more than twice the size of the Yellow River's. (*Appendix 5 – Yangtze River Drainage*)

The river is the lifeline of the Chinese economy because it supports a major part of China's agriculture, industry, and cities.[1] More than 400 million people depend directly on the Yangtze and its branches and tributaries for their livelihoods. The "endless" Yangtze also connects the Chinese to the past and, as a harbinger of things to come, to the future.

But it does not take much for anyone traveling up or down this great waterway to see that it is in trouble. One does not have to be a hydrologist or an environmentalist to see in an instant that the Yangtze's waters are just about as muddy and turgid as the

Yellow River's and that there is garbage and other waste everywhere.

However, when that pollution is stacked up against the torrential, destructive floods that hit periodically, it may be only a secondary concern to those who live alongside it. Still, the floods appear to be a rather recent phenomenon. If one examines records of more than two millennia ago, they show major floods striking the river every few decades on average. Over the past few centuries, however, as humans began to have a greater impact on the river and its drainage area, floods became more frequent and more destructive.[2]

In 1949, after it took over, the Chinese Communist Party right away started pumping money into taming the Yangtze. It raised about 3,600 kilometers of embankments on the main stream and more than 30,000 kilometers of levees by as much as 6 meters. For this, it used more than 4 billion cubic meters of dirt and stone, or enough material to put a wall around the globe three times.

In spite of this Herculean effort, it was hard to see any real benefit, especially after more floods hit. There was the great flood of 1954, the effects of which were felt almost everywhere along the Yangtze River basin, and there was a series of floods in 1980, 1981, 1983, 1991, and 1996. Then in June 1998 an immense flood upset the entire river valley. The reinforced embankments and levees proved to be practically worthless—9,000 of them collapsed in two months' time.

The central government gave the order, and 6.7 million laborers and more than 200,000 soldiers were sent off to man the embankments and do battle with the flood. They used their bare hands to build a secondary embankment on both sides of a 2,843-kilometer-long section in a desperate race against the rising floodwaters.[3]

Those efforts notwithstanding, the flood of '98 took the lives of 1,562 people, brought down 3.29 million houses, and did 134.5 billion yuan worth of damage. But many of the participants still felt victorious because the great river continued its eastward flow without causing a major breach—this during what was said to be one of the worst floods in history.

Was it really the worst, though? On one 359-kilometer stretch, the flood peak was a record. But in terms of volume, it was not a record at all as can be seen in Table 1.

TABLE 1 Peak Water Flows at the Yichang Hydrometric Station
(m³/sec)

Frequency (years)	1,000	500	100	50	20	10	5
Peak Flow	98,800	94,600	83,700	79,000	72,300	66,600	60,300

This can be compared to previous flood peaks at various hydrometric stations in Table 2 which looks at flood peaks during the two great floods of 1931 and 1954.

TABLE 2 Flood Peaks at Four Hydrometric Stations
(m³/sec)

Station	1998	1954	1931
Yichang	63,300	66,800	64,600
Luoshan	67,800	78,800	N/A
Hankou	71,100	76,100	59,900
Datong	82,300	92,600	N/A

In looking at these data, Chinese flood experts have pointed out that the volume of the 1998 flood was much smaller than that of two other large floods in the twentieth century alone. Many people have been puzzled by the fact that smaller floods occurring every six to eight years could break height records when they hit almost every middle Yangtze city along the way. And they are left wondering what will happen when the really big one comes along. After the 1998 flood, many Chinese were left asking: "What the hell is wrong with the Yangtze?"

The expression "great river flowing eastward" represents the role the Yangtze has played in Chinese geography, certainly for centuries if not for millennia. But why has China been forced to put so much effort year after year into blocking and taming the great waterway, and then, after celebrating a victory, have to turn around and worry about the next year and the year after that?

The answer to that question is another question: Just how much damage has been done to the mighty river, and how can it be corrected, if it can?

"Huguang's Harvests Could Feed All Under Heaven"

Numerous rhinoceros, deer, fish, turtles, and alligators thrive in Cloud Lake and Dream Lake, two huge wetlands between the Yangtze and its tributary the Han.
     —*Mo Zi* (468 B.C.–376 B.C.)

RETURN JUST FOR A MOMENT TO THE YELLOW RIVER. For ages, the Chinese have referred to it as the "mother river" because it was long believed that its valley was where China's glorious civilization began.

Recent discoveries from archeological digs seem to suggest that the Yangtze River valley may have been another cradle of Chinese civilization. But the economy and population of the Yangtze River valley remained much smaller than that of the Yellow River valley for centuries. That is one simple reason why the ecosystem of the Yangtze valley remained so sound while that of the Yellow River was destroyed long ago.

There were three large migrations from the Yellow River valley to the Yangtze River valley.[4] These human waves led to a considerable amount of development, mainly along the lower reaches and across the delta. That major site of resettlement, in areas on the lower reaches that are now Jiangsu, Zhejiang, Anhui, and Jiangxi provinces, began to be hit by more serious flooding at that time. But, generally speaking, the Yangtze River remained fairly unsullied long after the Yellow became "China's sorrow."

It was mainly the great wetlands of the Huguang Plain that absorbed the excess amount of floodwater. *Hu* in Chinese means "lake," *guang* is "a broad area." The *hu* refers to two bodies of water in the region that in ancient times were called Yun and Meng, or "cloud" and "dream." Their role in the Yangtze River system was to serve as catch basins whenever the great river was swollen with runoff from melting snow or torrential rainfall along the upper reaches. Throughout much of China's early history, these two lakes played a crucial role in preventing massive flooding and allowing people downstream to live in peace.

The two lakes began to shrink as early as the Han Dynasty (202 B.C.–A.D. 220), and Dongting Lake, just to the south, began serving as the main catch basin for Yangtze River floodwaters. During the flood season, Dongting Lake could expand to 6,000 square kilometers, making it the largest freshwater lake in China.

For centuries, attempts at reclamation were unsuccessful.[5] The Yangtze floods were so severe that any embankments that were rebuilt collapsed just as often, making it almost impossible to farm regularly in low-lying areas. So the vast wetland was left untouched as a natural flood basin until 500 years ago, when the two old lakes gradually shrank into a number of smaller lakes and the land north of the Jing River section of the Yangtze became gentle plains offering plenty of fertile soil for all takers — and there were to be plenty.

Population growth in the region quickly outstripped its environmental adaptability, and there was enormous pressure to increase the amount of fertile land by reclaiming river valleys and lake areas, of which there were many. Across the Huguang, in Hunan and Hubei provinces, the demand for land was so great that by the Ming Dynasty people were building dikes as if they were driven by demons. In 1542 the last floodwater-retaining area north of the river was filled in, turning the Jing River dike into one continuous earthen embankment for 124 kilometers.

After that, the Jianghan Plain north of the river became a land reclamation paradise that brought immense prosperity to Huguang.[6] Both Hunan and Hubei became major grain producers from the Ming on, leading people everywhere to declaim: "Huguang harvests could feed all under heaven." Unfortunately, all that embankment work and land reclamation set the stage for the disasters to come. It was a repetition of what had happened on the Yellow River centuries before.

And the floodwaters came, more frequently and more severely. That was because of the gradual loss of the many lakes and rivers of the Yangtze valley. The severity of the floods caused more people to suffer financially, and they, like their ancestors here and up north, had little choice but to flee and go seek their fortunes elsewhere.

In spite of these obviously troubling developments, the rulers of the Qing Dynasty failed to reverse the trend, instead encouraging even greater reclamation and continuing to allow the large-scale migration of people into the region.[7]

Pretty soon it became clear that Huguang could no longer "feed all under heaven." The increase in land reclamation was not able to keep up with the dramatic increase in population.[8] In no time at all, the two provinces went from areas with a low population density to just the opposite, and they had all the problems associated with overcrowding in a rural area.

But one of the worst of the new afflictions was the increase in the number of floods, which ended up forcing many of the happily resettled new residents living along the riverbanks to move up to hillier ground around the valley. For centuries, man had claimed the land that belonged to the lakes and rivers. That caused more flooding. When throngs of people began to move up to higher ground, they naturally began to cut more trees and farm the hillsides. That destroyed the fragile ecology. This went on unabated until the damage to the valley was irreversible.

SLASH AND BURN

Throngs of migrants from Hunan and Hubei moved into our mountains just to grow corn.
   —*Shiquan County Gazetter,* Shannxi province, 1850

THE WAVES OF MIGRATION INTO THE YANGTZE RIVER VALLEY followed a route that began in western Hunan and Hubei provinces and southern Shaanxi and headed westward into the forests of eastern Sichuan.

The practice at that time was to chop down trees in the spring, burn them, and get the ash for fertilizer just before the onslaught of the rainy season. When the rains ceased and the land was still warm, all the underbrush would be cleared and planting would begin. This usually led to a great harvest.

Unfortunately, this was the Chinese version of slash-and-burn agriculture. Tree by tree, the ancient, primeval forests that had taken hundreds or thousands of years to grow and which had protected the hills and mountains fell in a reckless pattern. But nature, even if it does at times appear passive, got revenge.

By the mid-Ming Dynasty (sixteenth century), corn, peanuts, sweet potatoes, and tobacco had been introduced into China from the New World and began to play a part in the destruction of the forest cover along the upper and middle reaches of the Yangtze. Corn and potatoes, with their ability to withstand temperature variations and unfriendly soil, became two of the more lethal weapons in the fight against the forests. Previously there had been no food crops hardy enough to withstand the cold and the unfertile soil of the higher elevations.

Throughout the region, corn and other crops gradually replaced the great primeval forests, but the prosperity was short-lived. The large-scale felling and burning of trees caused a loosening of the soil, and when the rains came, it joined the ash fertilizer in being easily washed away. The local people had only one choice, and that was to move once again—deeper into the mountains.[9]

The mountainous parts of western Hunan and Hubei were primarily occupied by the Miao ethnic group. Their patriarchs, according to the policy of the time, had been given sole authority over the area. But by the time of the Qing's Yongzheng Emperor (1723–36), these rights were gradually being rescinded and replaced by a prefecture that was centrally administered.[10] An army of immigrants was now given the right to go to the region and engage in agriculture.

Western Hubei and its mountainous areas were always a site of struggle for newly arrived immigrants, who had unleashed attacks on the forests for centuries. From 1711 to 1784, in three prefectures previously run by the Miao, the population increased by 3.8 percent annually. This meant more pressure on the land as forests gave way to cultivation and the environment gave way to everything.

When the mountainous parts of western Hunan and Hubei could no longer stand the population strain, people were forced to move farther west, deeper into the mountains on the upper reaches of the Yangtze, and then fan out along its tributaries, sometimes to the south. The forests of Sichuan, Shanxi, Guizhou, and Yunnan, which had provided an ecological barrier, were gradually replaced by waves of immigrants and hardy crops.

Over time this inexorable cycle led to the degradation of the Yangtze. Lakes were drained to provide arable land, which led to a population explosion. That led to a struggle to reclaim even more land from more lakes and the river itself, causing more flooding along the entire river basin. That in turn caused a migration to the hillier and more mountainous areas, where it was the forests and vegetation that were sacrificed in the struggle to make more tillable ground. The settlers triumphed, in a manner of speaking, but serious soil erosion problems followed, which led to a dramatic increase in the amount of sediment washed into the Yangtze and nearby lakes. This in turn led to another round of reclamation of lake areas, and so on and so on.

From the end of China's last dynasty, the Qing, in 1911 through the Nationalist period (1912–49), this cycle left the people and their surroundings increasingly desperate; eventually any attempt to correct it was bound to be pointless.

The destruction to the Yangtze during the last period of China's dynastic rule was a turning point. It went from a generally benign and harmless waterway to an enormously destructive one. The frequency of flooding in the densely populated plains of Hunan and Hubei became in many ways China's number one headache—a migraine, in fact, for which no pain reliever has been found to this day.

"IF YOU DON'T DIE, I WILL"

Hunanese regard their dikes as lifesavers, while their Hubei neighbors consider them a lethal weapon.
—Zhang Zhidong, governor of Hunan and Hubei, 1892

AT THE MOST CRUCIAL POINT IN THE 1998 FLOOD ORDEAL, thousands of people were working day and night on the dikes that protected villages and towns. But, strange to behold, when a dike finally gave way, the people on a neighboring dike within sight, laboring hard to stop the flood, would actually start cheering, although the breach would mean of course that their neighbors' property and everything else they owned was washed away.

The reader may well wonder about this odd behavior and people finding satisfaction in the suffering of others, but the fact of the matter was that people living in these areas of reclaimed land knew that unless other dikes gave in and acted as a relief valve by letting water flow into the old catch basins, the floodwaters would overrun their own land. This was merely the repetition of an old idea: "If you don't die, I will."

This was the thinking that lay behind the resentment that had smoldered for generations between the people of Hubei and Hunan. They were separated by the lofty embankments of the Yangtze River — but by very little else.

In general, the Jianghan Plain in Hubei (the province's name means "north of the lake") and Dongting Lake in Hunan (meaning "south of the lake") were two immense low-lying areas protected by equally immense levees. A 337-kilometer section of water referred to as the Jing River crossed the lowest area between the two and had no real fixed course. It simply ran over into the big lakes during the flood season. But this traditional function gradually changed because of dike and embankment building, which turned it into a suspended, single-course waterway that was eventually the cause of a conflict between Hubei and Hunan.

As noted above, the land reclamation around the two lakes was excessive and rarely, if ever, coordinated. The embankment work had been going on much longer in Hubei than in Hunan. The Jing River dike referred only to embankments and levees on the northern bank, that is, the Hubei side. Prior to the Ming, interprovince conflicts were held in check by sluice gates that had been dug into both the southern and northern banks of the river and were used to divert any floodwater.

But embankments and levees were raised throughout the Ming, and by 1542 many of the sluice gates had obstructions. Continued building of dikes on the northern side allowed floodwaters to spill over onto Hunan's plains, putting its population at risk.

In 1650, during the mid-Qing, the last of the northern sluice gates was permanently blocked, making year-round cultivation of the Jianghan Plain possible, but only by putting an end to the gate's original role. The entire burden of taking in any increased amount of water fell to Dongting Lake, on the south side. The population around Dongting was growing rapidly, and that cut into the lake area, reducing its ability to hold floodwater. The flood threat in the lake area increased.[11]

This threat ultimately caused people on both sides of the Yangtze to demand that the big river be tamed. But that was about as far as any agreement went and where any similarity in demands ended. The folks in Hubei demanded that Dongting be restored to its previous size by giving up the reclaimed land *(feitian huanhu)* because, they said, it was Hunan that had reduced the size of the Dongting catch basin and the amount of flood storage.

The folks of Hunan replied that they wanted the river restored by blocking their sluice gates *(saikou huanjiang)* because originally the diverted floodwaters and Jing runoff were shared evenly. But when the sluice gates on the northern (Hubei) side had been blocked, virtually all floodwater went southward, which, the people in Hunan pointed out, was grossly unfair. The best solution, they suggested, would be one where that status quo was restored and both sides once again bore the flood burden, or, failing that, the southern sluice gates should be blocked to reduce the burden.

That debate raged on with no end in sight. Meanwhile the destruction of the forests along the upper reaches of the Yangtze continued unabated and Dongting Lake continued to take in silt, leading to more land reclamation and a further reduction in the lake's size.

Because Hunan and Hubei could not reach a settlement, the conflict smoldered until it burst into flame during the Nationalist era (1912–49) and the government had to step in. Once again it favored the Hubei side and ordered Hunan to return the farmland to the lake. The Hunanese were less than scrupulous in carrying out that order.

It may be difficult to look at the situation objectively and say which proposition would have been sounder. Hubei's position that the lake area needed to be restored to permit Dongting to hold more water and take pressure off the Jing River dike seemed reasonable enough—until one considers that the increased levels of sediment meant that the lake's ability to store floodwaters was being compromised in another way.

Also, the area around Dongting was the core of Hunan's agricultural area, and implementation of the Hubei proposal would have inundated much of this land and led to a massive forced migration of destitute people.

On the other hand, Hunan's proposal that the sluice gates be blocked would have reduced the amount of floodwater going south but resulted in a 30 percent increase in the amount of water flowing north into the Jing section. In all likelihood, the drastic change could have caused the already precarious Jing River dike to collapse altogether, resulting in enormous losses in the heavily populated Jianghan Plain.

The sediment that would no longer be accumulating in Dongting would then make its way into the Jing, adding to the sediment that had already reduced the river's natural flow and elevated its bed.

Apparently there was no equitable approach, one that would benefit both of the contentious provinces, unless a way could be found to stop the large amount of sedimentation caused by deforestation. For centuries, no government would dare get close to this sensible, but highly undesirable, long-term solution.

IMPOUNDING FLOODWATER AND RECLAIMING LAND

When numerous lakes were filled up for farming in the decades of struggle between the people and water for low-lying land along the Yangtze River, the retention capacity was much reduced and the flood situation exacerbated.
    —*China's Great Flood of 1998*, Water Resources Ministry

CHINA DEALT WITH THE YANGTZE just as it had with the Yellow River, by using an immense amount of manpower and other resources to try to tame it. Worries about how to find food for its growing population while increasing industrial production kept it from looking for concrete ways to restore its frayed ecosystem.

One of the more intractable problems was that when one considered the population problem, the practice of taking land from lakes, and the slash-and-burn techniques, the "new China" bore an alarming resemblance to the "old China" and the sins of the fathers were visited upon the descendants for generations.[12]

In the early 1950s, the Yangtze River Water Resources Commission (Changjiang shuili weiyuanhui) was established. It immediately set about tackling the problem on the Hunan and Hubei plains. From the start, the commission members ignored the various proposals made by Hunan and Hubei and came up with their own approach, which at first appeared to be rather clever.

This was the "impounding floodwater and reclaiming land" (*xuhong kenzhi*) method. It was straightforward enough. The basic idea was to construct a series of floodgates at the entry point of just about every lake feeding into the Yangtze so that whenever serious flooding hit the main stream, the overflow could be directed into these gates and on into lakes and newly constructed basins. In the case of relatively minor floods, the dikes and embankments would be the sole means of flood control, making the lakes and basins the rear guard and final line of defense against torrential waters from a major flood. Low-water levels would accord with the design and capacity of catch basins to ensure that they would be available when the really big one came along. Then, during periods of normal river water flow—and, more importantly, during the dry season, when the water in lakes and basins receded—the exposed land could be used for growing crops.

In addition to the lakes, new catch basins would be built with that same dual purpose. This would allow anywhere from 3 billion to 6 billion cubic meters of floodwater to be diverted when necessary, while at other times of the year their barren spots would provide additional fertile land for cultivation.[13]

In theory, catching floodwater occasionally and gaining land for the rest of the time looked like a winner no matter what. Once again, however, reality ran way ahead of the plan, which in any case just became a good excuse for people to reclaim lake land anytime they felt like it. It eventually led to a horrible reduction in the amount of catch basin area.

Today, in hindsight, that winning policy can be seen as the failure it was. People lived successfully for a long time in those homemade catch basins, the ones that saw practically no floods. They went on to marry and build houses, roads, and schools there.[14] It is hardly surprising, then, that when the monster floods hit in the summer of 1998, no one had the heart to open those floodgates. It would have destroyed those people's lives.[15]

The initial success of the man-made catch basins was good enough for those people, who were eager to get their hands on lake land they could use for farming. An astonishing number of lakes on the Jianghan Plain, the original catch basins, were filled in completely. There were 1,066 major lakes on the Jianghan Plain in the early 1950s. That number shrank to 629 by the end of the decade. By the early 1980s it was a mere 309. *(Appendix 6 – Lake Reclamation on the Jianghan Plain)*

Hubei had allowed so many of its lakes to be reclaimed for agriculture that it shifted the entire burden of flood control to the southern side, Hunan. This meant Dongting Lake almost exclusively. In response to protests from Hunan, in 1958 the central government finally blocked off one of the four outlets south of the river.[16] The three remaining outlets, unfortunately, saw to it that the sedimentation buildup in Dongting continued — as did the practice of reclaiming land in the lake area.[17]

At the same time, there was a rather bizarre engineering project in the works whose purpose was to take some of the kinks out of the Jing River section of the Yangtze.

In all, there were three such projects started during that period. In the end the Jing lost about 78 kilometers of its original length. The benefits of these costly projects were dubious, to say the least.[18]

In the past two decades both Hunan and Hubei have seen severe flooding a number of times, which indicates that all of the clever engineering ploys described above had little or no noticeable effect on the colossal tantrums of the ill-behaved Yangtze. Those very elaborate efforts began with a comprehensive plan to arrange floodwater diversions and provide reclaimed land. In the end they simply failed.

Obviously, some other scheme would have to be dreamed up. And when it was, many people believed that had finally found the solution: a huge dam, a really big one, a dam to end all dams.

SMOOTH LAKE IN HIGH NARROW GORGES

Great plans are afoot . . .
Walls of stone will stand upstream to the west
To hold back Wushan Mountain's clouds and rain
Till a smooth lake rises in the narrow gorges
The mountain goddess, if she is still there
Will marvel at a world so changed.
　　　　—"Swimming," Mao Zedong, 1956

THE PROPOSAL WAS SIMPLE. Because the Yangtze valley covers more than 1.8 million square kilometers and rainfall there is estimated at around 960 billion cubic meters annually, a great big something or a great deal of something would be needed. And there was a big something, a big dream that both Chinese and foreigners had been having for a long, long time—the Three Gorges Dam.[19] Yes, build a massive dam on the Yangtze River, at Sandouping, in the Three Gorges.

For many, it is the only possible way to tame the mighty river. This was believed to be a truly comprehensive plan, one that would bring not only better flood control but also electricity, irrigation, transportation, and fish.

Since 1949, China has built 48,000 large, medium-sized, or small reservoirs along the Yangtze, with a total storage capacity of 122 billion cubic meters. That number is thirteen times the size of those built on the Yellow River. But why have so many structures failed in their basic role of bringing the Yangtze River under control?

Supporters of the Three Gorges Dam project said that the answer was as plain as it was simple: Those 48,000 dams, with the exception of the Gezhou Dam, were built on the Yangtze's many branches and tributaries; not a single one was built on the main stream.

Let's go back briefly to 1949, the year "new China" was established. The Yangtze was hit by a large flood. Although the Jing River dike held, Dongting Lake practically became an ocean. In a flash, the new government got the brilliant idea of taming the Yangtze. It put that at the top of its agenda.

There is a story from that period about Mao Zedong's first visit to the Yangtze that bears a remarkable resemblance to the one about his first inspection of the Yellow River. No matter how apocryphal it might be, the results were exactly the same.

On February 19, 1953, Mao conducted his first inspection of the Yangtze. He was already dissatisfied with the rather lackluster plans or proposals for a dam and reservoir that were espoused by Lin Yishan, who was director of the Yangtze River Water Resources Commission. Mao wanted to know why there was no plan afoot for building a dam at the Three Gorges, one that could put an end to flooding at once. So, just as quickly as that, plans got under way for a dam that would indeed do just what Mao would later describe in his poem "Swimming," namely, a smooth lake rising in the narrow gorges.

Those were deeply troubling times, however, what with the Korean War and the attempt to build an entirely new China at the same time. By 1958, after several years of stops and starts, preparations were finally able to get under way in August, after a group of Soviet experts informed Premier Zhou Enlai that everything was ready. But fate was unkind, and within two years Sino-Soviet relations had deteriorated shockingly. Once again the project stopped dead in its tracks and attention turned to the much smaller Gezhouba Dam downstream, a kind of test case for the behemoth at the Three Gorges.[20]

It was not until almost three decades later, in 1984, that the central government's State Council got around to approving a plan for a dam at the Three Gorges that would raise the water level behind it to 150 meters above sea level. Almost immediately it became obvious that the amount of sediment that would or could build up behind such a dam could pose a danger way upstream to the city of Chongqing. So the proposal was shelved, and debate began anew over whether such a reservoir should ever be built.

A year later, the Chinese People's Political Consultative Conference's Economic Construction Committee organized a tour of Sichuan and Hubei provinces that lasted for thirty-eight days. Their investigation resulted in a report that was delivered to the central government. Its title: "The Three Gorges Project Should Not Go Ahead in the Short Term." It gave seven important reasons for the negative response:

1. Cost. *Total cost would be not 20 billion yuan (U.S. $2.5 billion), as originally thought, but 60 billion yuan.*

2. Flood control. *Not only would this problem not be solved on the middle and lower reaches, the dam would actually increase the possibility of flooding on the upper reaches.*

3. Silt and sediment buildup. *Couldn't be resolved.*

4. Navigation. *More harm than good.*

5. Electric energy output. *High cost, long construction period, slow output, poor results.*

6. Population resettlement. *More than a dozen cities would need to be moved.*

7. Safety. *Potential disaster in case of a nuclear attack.*

The State Council responded by requesting a complete reassessment of the project by the Ministry of Water Resources and Electrical Power.

In November 1986 the ministry got 412 experts and 21 other specialists of various kinds and formed them into fourteen separate panels to analyze the problems and debate the pros and cons of the giant project over a two-year period. The conclusion this time: "Better to build the dam than not. And the sooner the better."

In 1992, the National People's Congress (China's nominal parliament) voted on the project. The results: 1,767 in favor, 177 opposed, and 644 abstaining. And with that, the most highly controversial and hotly contended proposal in the history of the PRC was approved and the seventy-year old dream began to take shape.

But it was not the original plan. The new one emphasized the project's multifaceted nature: not just electricity, but also transportation, subordinated, of course, to the primary function— flood control.

Control of the devastating floods is what makes people support the dam despite the huge ecological damage and all the sufferings of over 1 million relocatees. But while the people of Hunan and Hubei and their perennial flood problems must have been a deciding factor in getting the go-ahead, it wasn't long before the participants began to wake up to the fact that just one structure was unlikely to solve such a complex problem; in fact, it was highly unrealistic.

The upper reaches of the Yangtze have an annual flow of 451 billion cubic meters. The flood control capacity of the dam's entire 600-kilometer-long reservoir is only 22 billion cubic meters, while total storage is set at 39.3 billion cubic meters. In view of this, water levels in the reservoir should be kept very low during flood season to make sure that it can handle a large amount of floodwater. It means that lesser floods will have to be allowed to flow downstream. But there's a problem. With Dongting Lake so badly silted up and its capacity reduced, even small floods could lead to big disasters.

One might consider what happened in the relatively small floods of 1980 and 1983, which caused a big overflow in the lake. Then, in 1996, a disaster hit when rising waters spilled over into reclaimed parts of the lake, killing 170 people, leaving more than a million homeless, and doing more than 30 billion yuan in damage. The water flow that was measured at the Yichang Hydrometric Station on the upper reaches during these floods would be considered too small an amount to bother holding in the Three Gorges during flood season, when levels would need to be kept very low to prepare for the big one. Small floods aren't considered to count, but they can do big damage downstream.

The Three Gorges Dam project needs to be approached carefully because of changes that could take place in the delicate balance between the Jing River and Dongting Lake.[21] When the Three Gorges Dam is completed and large amounts of sediment begin to flow into the reservoir, the water going downstream will have less sediment and be clearer. This should have a secondary effect of removing sediment on the riverbed downstream. In theory, that should benefit both the Yangtze and Dongting Lake. But researchers have found that sediment deposited at the river-lake junction can remain there for twenty years before being carried farther downstream. That means worse floods for Dongting Lake.

There is another potential problem upstream for navigation. The area around the port of Chongqing can be affected by the sediment backing up from the reservoir. To deal with this problem, the government has approved two more dams on the Jinsha, as the Yangtze is known on its upper reaches: the Xiluodu and Xiangjiaba, simply to stop sediment. The first will cost 110 billion yuan.

Even this massive engineering feat appears to be exactly like all previous attempts—one that simply cannot solve the sediment problem. How can this damned sediment problem have become so bad over the past five decades?

"THE ENDLESS FALLING OF TREES"

The trees of our Ganmei are being chopped down and sent to float in the river night and day as if they were nothing but a bunch of noodles.
—He Tao, deputy director, Ganmei Tibetan Autonomous Prefecture, Sichuan province

THE ECOLOGICAL DESTRUCTION CYCLE began in Hunan and Hubei

provinces with a seemingly endless stream of people migrating to the mountainous western parts. That was followed by a move to eastern Sichuan, where they began using slash-and-burn techniques. As early as the Ming and certainly as late as the Qing, there were large strips of badly eroded land and a heavy increase in the sediment reaching nearby streams, rivers, and lakes—and ultimately the Yangtze.

This wild behavior continued for several centuries, and no government—whether imperial or Nationalist—was able to stop it. Yet by the early 1950s there were still large densely forested areas high in the mountains of western Sichuan and northwestern Yunnan. These valuable reserves should not have been touched.

But that was not foremost in the government's mind at the time. So in the early 1950s those final forests started falling under the axe. One of these areas was along the upper reaches of the Min River—one of the Yangtze's most important tributaries. The Chinese government at that time established the Western Sichuan Forestry Bureau. One purpose of that innocuous-sounding bureaucratic office was to run the country's forestry industry. It operated in Aba (Ngawa in Tibetan), Ganmei (Banzi in Tibetan), and Liangshan Yi autonomous prefectures. There was an orgy of cutting, and the primeval forests along the Min, Dadu, Yalong, Jinsha, and Jialing rivers were gone.

Why? The country badly needed timber. When it came to cutting down old forests, the prevailing wisdom was "If primeval forests are not chopped down, the trees will fall on their own" and "Cutting old growth makes way for new." At least half of the forests of western Sichuan disappeared almost overnight.

In the 1960s, with the souring of relations with the USSR, China, then in a siege mentality, developed a fallback military defense strategy. Part of this was the "third front," an area of munitions factories and other industrial projects safely ensconced deep in the interior of China, like the steel facility at Panzhihua. Even more forests were sacrificed.

Lin Biao, the minister of defense and heir apparent to Mao, issued his well-known order calling for people to move out of urban areas as part of this schizophrenic plan. More than 23,000 people went helter-skelter to previously unpopulated forest regions of western Sichuan. Apart from a few border skirmishes, the war with the Soviets never came. Russian bombers did not fill the skies over China, but a random observer might have been excused for thinking that the area had been bombed. For more than a decade large tracts of forest were laid bare and cut with gashes of erosion, and it was the Yangtze River that paid the price.

Mao Zedong's death a decade later brought changes and the inauguration of economic reforms in 1978. The people in Sichuan began to calm down a bit and cooler heads prevailed. But when they awoke from the narcotic haze of the past, they discovered to their great horror that the dense forests that had covered 40 percent of the western part of their province now covered less than 20 percent of it. The province as a whole had only 12 percent coverage, and in twenty counties it was as low as 1 percent.

Initially, some of the folks in Sichuan thought they had lucked out. But in July 1981 fortune played an ugly hand and turned on them. Massive flooding hit the Chengdu Plain. The city of Chengdu, Sichuan's capital, is near the eastern foothills of those western mountains. The plain stretches far to the east to the immense city of Chongqing. The floods devastated the upper and lower reaches of the Min, Tuo, and Jialing rivers.[22]

Confronted with this ruinous state of affairs, Sichuan began to eat itself up with worry. So in October 1981, the China Forestry Society organized a group of thirty experts to take a close look at the situation. They promptly delivered a report, "Protecting Forests Along the Headwaters Is the Fundamental Goal of Taming the Yangtze River: Memorandum of an Investigation of Soil Conservation on the Yangtze River Valley." This led to a major campaign to plant trees in the region, which resulted in a marginal increase in the forest cover on the plain.

Young trees have very little impact when it comes to soil conservation, so the Yangtze River valley derived little benefit initially. But that was not the worst. This putative increase in the amount of coverage gave people the wrong impression: They thought it was okay just to keep on doing what they had been doing for so long, and they continued chopping down the natural forests. Before 1985, there had been 3.6 million hectares of natural forests in western Sichuan. By 1995, that was down to 2.34 million.

But laying all the blame on state-sponsored loggers would be a bit unfair. Before the 1980s, virtually all forests were state-owned and the deforestation was the sole responsibility of state-run companies. That somewhat irritated the local people, who saw few rewards from the loss of their treasured resources but bore much of the brunt if things went wrong.

But in 1982 the administration of some forests was handed over to county governments, who in turn designated them prime timber sources and allowed county-run enterprises to cut them down. The remaining forests were turned over to townships, villages, or even private entrepreneurs, who immediately set about turning large amounts of wood into charcoal for home cooking and heating. This situation became known as the "five axes" — referring to the state, county, township, village, and individual — all competing for the same dwindling numbers of trees.

Another important factor in the increased deforestation was technological change. In the 1950s and 1960s logging companies were relatively backward and could only manage a few trees a day with their axes, crosscut saws, and other relatively inefficient tools. Then the chain saw arrived and the labor-intensive work was reduced to the point where one man could fell a tree in less than ten minutes. In just one day a semiskilled worker could bring down upward of 300 trees. Transport was another important element in the industry. The wood was useless if it couldn't reach markets well to the east. But modernity changed that as well.

By the 1990s Sichuan was flooded with private entrepreneurs setting up mostly unregulated timber companies. There were something like 10,000 in western Sichuan alone. When it was possible, the timber was skidded down the mountainsides or hills into tributaries, where it was lashed together so that it could be floated downstream. The logging took away the forests of Ganmei and Liangshan, two regions along the upper reaches.[23]

Then there was Aba Autonomous Prefecture, on a plateau area of northwestern Sichuan, near the upper reaches of the Dadu and Min rivers. Its forest coverage had been the highest in all of Sichuan. After 1949, it provided more than 70 million cubic meters of timber for state-owned companies, and forest coverage went from 34.4 percent in the 1950s to only 18 percent in the 1980s. This obviously could not continue much longer.

TABLE 3 Timber Production in Western Sichuan
and Eastern Yunnan Provinces (000xM³)

Province	1957	1965	1978	1980	1985	1986	1987
Sichuan	1,499	1,986	3,756	4,157	4,224	4,253	4,316
Yunnan	507	1,126	2,117	2,456	3,315	3,413	3,560

There were forest workers in western Sichuan who felt a tinge of regret for having cut down so many trees, especially when they saw the floods of 1998. One of them was Tang Song, an Aba lumberman who had been given a National Model Worker award for felling 30,000 trees in twenty years. He openly expressed his sense of guilt when many towns in the prefecture were destroyed by mudflows in 1998. But the damage was done and the regret came a bit too late. Tang didn't have a choice, in any case—he'd lost his job because his company had completely laid waste to the forests in the surrounding mountains.

A SECOND YELLOW RIVER

In the winter and spring, the Yangtze River is so crystal clear at the Three Gorges that it reflects everything above it like a huge mirror.
—*Reference to the Book of Rivers*, Li Daoyuan (A.D. 466–527)

"CALLING YANGTZE, YANGTZE, THIS IS YELLOW RIVER, do you read me, over?"

"Receiving you, Yellow . . . This is also the Yellow River, over."

That exchange above is a bit of tongue-in-cheek humor from the magazine *Satire and Humor*. It's delivered in that mock military communications language often heard in war movies and is meant to poke fun at the two rivers. The joke is based on something that's not just some laughable coincidence. The Yangtze has had an alarming increase in the amount of sediment it carries, and a cogent argument can be made that it is like a second Yellow River. In addition, the process of turning the Yangtze into a second Yellow River was related to the loss of tree cover, especially in older-growth areas.[24]

The most serious deforestation is in Aba Autonomous Prefecture, which has serious soil erosion problems on almost 60 percent of its land. It has become a sort of playground for mudslides and other terrestrial disasters. In the 1950s, there were only two serious ones; then in the 1960s it was four, and in the 1980s more than a dozen.

There are soil erosion problems on 44 percent of the entire Min River valley. The upper Min now gets more than 100 million tons of sediment annually, and a thousand serious mudslides have been recorded. Average sediment levels on the Min and its tributaries have doubled or in some cases even tripled.

Similar problems can be found along the Dadu, the hillsides of which in recent years have become completely barren. The river had serious soil erosion problems on a 12,000-square-kilometer area in the 1950s, but by 1992 that had expanded to 19,846 square kilometers. That area produced 50 million tons of sediment annually on average for the Yangtze and its tributaries. That is 10 percent of the Yangtze's sediment on its upper reaches.

Things in Ganmei Autonomous Prefecture are no better. Virtually all the forests on the western bank of the Dadu have been cut down, and the river gets 100 million tons of sediment annually, with half settling on the riverbed or in reservoirs, the other half going into the Yangtze.

Then there is Liangshan Autonomous Prefecture, at the Yalong-Jinsha river junction. The steep mountains and deep valleys of the region have been hit badly by mudslides during normal rainy periods, meaning a substantial increase in the amount of sediment in the Yalong and a silting up of the Yangtze.

And in Yunnan province, to the south of Sichuan, the situation has been just as bad. In fact, the deforestation has gotten even worse. Yunnan's deep valleys have become a major source of mud for the Jinsha.

The Longchuan, a tributary of the Jinsha, has been the recipient of enormous amounts of eroded soil. In 1993, erosion affected a 4,500-square-kilometer part of the river valley. After that, the Longchuan, which is a mere 230 kilometers long, dumps 6.2 million tons of sediment into the Jinsha every year.

And finally there is Yunnan's Dongchuan prefecture, where there is a small branch of the Jinsha a mere 132 kilometers long. It has practically become a mudslide museum—there were 50 during the 1950s, and in recent years twice that, 107. On seven different occasions in the past fifty years, the river's flow was blocked by a mass of mud. The annual sediment running into the river has been estimated at an astonishing 40 million tons, 6 million of which makes its way into the upper Yangtze. One expert has pointed out that if the 107 mudslides had all occurred at once, the city of Dongchuan would have been obliterated in five minutes—a muddy Pompeii.

The joint Sichuan-Yunnan "effort" in the Jinsha River valley has made it one of the more severely eroded parts of the Yangtze drainage area, with more than half of the entire valley given over to soil erosion. Along both sides of the Jinsha's 2,308 kilometers there have been more than 1,100 mudslides. Those mudslides and collapsing embankments have turned it into the Yangtze's greatest source of sediment. Annual sedimentation amounts to 280 million tons on average.

Soil erosion was unavoidable along the upper reaches after the deforestation began. Primeval forests gave way to secondary forests; those gave way to brush, then grassy slopes, and eventually barren hills. The ecology was destroyed.[25]

Central government policies consistently abetted the deforestation process. That was especially true during the big campaigns of the 1950s and 1960s. During the Great Leap Forward (1958–60) there was a push to increase China's steel production through the use of so-called backyard furnaces. It was that bizarre campaign that had everyone donating every piece of scrap iron, or in many cases iron that wasn't scrap, to the community smelter so that every pot, pan, bedstead, or whatever else was available could be melted down to produce iron locally. The result was the most appalling collection of worthless pig iron one could possibly imagine.

This campaign led to a massive felling of trees to supply the wood that would stoke the homemade furnaces that dotted the countryside and the cities. The felling of trees aggravated the widespread drought in Sichuan in the early 1960s and led to a sharp drop in grain production over a three-year period and a corresponding drop in the population of 3.6 million people. The disastrous backyard furnaces were soon shut down, but the hillside farming did not stop and there was irreparable damage from deforestation. Sichuan province began contributing 680 million tons of sediment to the Yangtze annually.

To the north, there are the Bailong and Baishui rivers of southern Gansu on the upper reaches of the Jialing River, a major tributary of the Yangtze. In 1984 southern Gansu had 29,000 fewer hectares of forest than it did in the 1950s.

The situation is no different on the southern slopes of the Qin mountains, to the east in southern Shaanxi, on the upper reaches of the Han and Jialing rivers, which cover nearly 80,000 square kilometers of ground. These denuded areas contributed an enormous amount of soil to the Yangtze.[26]

Finally, there is Shennongjia, at the juncture of Sichuan and Hubei provinces. It was known for its primeval forests and rare plants. It had 300,000 hectares of forest in 1958, but by the 1980s only 10,000 hectares were left.

TABLE 4 Percent Change in Forest Cover in
the Three Gorges Area (%)

County	1950s	1980s	County	1950s	1980s
Changshou	18.5	7.5	Yunyang	–	8.91
Fuling	–	11.53	Fengjie	32.3	15.7
Fengdu	23.7	12.95	Wushan	23.6	11.7
Shizhu	23.3	10.97	Wuxi	24	10.5
Wulong	–	13.02	Badong	–	32.5
Zhongxian	22.2	11.6	Zigui	–	25.9
Wanxian	20	10.2	Xingshan	–	–
Kaixian	11	5.9	Yichang	–	36

The 82,000-square-kilometer municipal district of Chongqing (which is a provincial-level city like Beijing, Shanghai, and Tianjin) was set up to help deal with the Three Gorges Dam project and the population shift. But 52 percent of the hilly terrain around it has serious soil erosion problems, producing 200 million tons of sediment annually, 140 million tons of which ends up in the Yangtze.

In a few short decades, the once relatively clear waters of the Yangtze and its major tributaries turned murky and became a kind of second Yellow River. The waters at the source were more choked with sediment, so how could the Yangtze possibly have been expected to escape the same fate?

"WATER GOES, ROCKS APPEAR"

Farmers had to move and open new lands elsewhere after three harvests on one piece of hillside ground in western Hubei because the 10 centimeters of soil would be completely washed away in three years, leaving nothing but hard rock.
    —*Tour of Rongmei*, Gu Cai, 1703

THE QUESTION OF WHETHER THERE IS A SECOND YELLOW RIVER or not has spawned a considerable amount of debate, but it is mostly confined to those who believe that the process is well under way versus those who argue that the likelihood of the Yangtze becoming like its northern neighbor is exaggerated.

Both groups have a mountain of data to back up their claims. There is one area where the two sides generally agree, which is that the soil erosion on the upper reaches has been awful and that the severely eroded area has grown to 550,000 square kilometers.

Those who believe that the Yangtze is already a second Yellow River point to the following:

Fact: Even though the waters of the Yangtze and its branches and tributaries have lower sediment concentrations than the Yellow, the 2.4 billion tons of sediment flowing into the Yangtze amount to more than the 1.6 billion tons of eroded soil going into the Yellow.

Fact: In the 1980s and first half of the 1990s, the average sediment levels in the main part of the Yangtze passed the 700-million-ton mark. That's more than the sediment in the Nile, Amazon, and Mississippi put together. The problem is getting worse.

Fact: Reservoirs are worthless in solving the sediment problem because of the soil erosion along the upper reaches of the Yangtze.[27]

Those people who believe that the Yangtze is not even close to becoming a second Yellow River have marshaled their own impressive array of facts. They believe the Yangtze is still qualitatively different and point to the following:

Fact: In spite of the agreement on the amount of land with significant soil erosion problems, the soil in the Yangtze drainage area is much coarser than the loess soil of the Yellow River. Much of the eroded soil never reaches the main body of the Yangtze, especially since the tributaries and their outlets into the Yangtze are relatively low-lying and slow down the flow.

Fact: In the latter half of the 1980s, the water level in the Yangtze dropped, as did average sediment levels. This should lead to a complete reevaluation of the sediment problem.

Fact: The sediment buildup and other problems do, however, mean that more reservoirs and water control projects are needed. According to the assessment of the Three Gorges Dam project, the more than 11,000 reservoirs on the Yangtze's upper reaches play a crucial role in stopping 230 million tons of sediment annually and in reducing sediment in the middle and lower reaches. The fact that the riverbed in the lower reaches is gradually rising proves only that there are too few rather than too many dams.

Listening to both sides of the argument one might make one pause for just a moment. Both agree that 2.4 billion tons of soil are running into the Yangtze annually, 1.8 billion of that going into the upper reaches. A sensible question, then, is whether the upper reaches can withstand such enormous pressure.

Again, both sides generally agree that 70 percent of the land along the upper reaches faces severe erosion. The topsoil in the area is thin or scarce. On more than 13 million hectares of hillside land along the upper and middle reaches, the topsoil is less than

50 centimeters thick. The long period of deforestation that began as early as the Ming has, over the centuries, caused soil erosion and in some spots left only stone with little trace of fertile ground. Rocky ground with thin topsoil is the most common feature.

It takes an incredible amount of time to turn stone into soil. Just to produce 10 centimeters of soil, a 3-meter-thick piece of limestone is needed. The process of producing even 1 centimeter of earth can take as much as 10,000 years.

Now, turn that around and undo the process. Soil can be eroded about ten thousand times faster. If something is not done fairly soon, in less than 100 years the vast area of the upper reaches of the Yangtze will be nothing but rocky terrain unsuitable for human habitation. But people are scheduled to be moved away from the river and up into the mountains as part of the Three Gorges Dam resettlement project.

There are counties in the upper reaches of the Yangtze's drainage area that lose almost all their topsoil during the rainy season. Leaders there have to organize labor gangs to carry the soil back up to the hillsides from down in the river valleys just to farm.

As the territory along the upper reaches becomes rockier, the floods will only get worse until they are completely out of control. It should be clear enough from this that those who believe the Yangtze is becoming a second Yellow River are fairly close to the truth.

THE SOURCE OF ALL WATERS

Sing our praises for the Yangtze River, its bounty is endless.
    —*Song of the Yangtze*, Hu Hongwei (1984)

SO THE SECOND YELLOW RIVER DEBATE confines itself almost exclusively to the sediment buildup in the tributaries and the main stream. And the Yangtze's annual flow, which is twenty times that of the Yellow, gives some people a certain amount of reassurance.

But how can they be so confident that the Yangtze will really never have the same problems? Especially after a scholar in the megacity of Chongqing did a considerable amount of research and came up with the shocking prediction that, after only twenty years, the Yangtze would also have dry periods and would stop completely before reaching the sea near Shanghai.

What did he base his conclusion on? First, there were the changes already occurring on the upper reaches, especially on the Min, which is 735 kilometers long. Long ago it was thought to be the true headwaters of the Yangtze. It has an annual flow of 86.8 billion cubic meters — far greater than the Yellow.

As early as the Qin Dynasty (221–209 B.C.), there was a large diversion project at Dujiangyan. It was built by Li Bing and his son at a strategic junction, where the Min spills out onto the Chengdu Plain. For generations people looked to the Min for their livelihood. Now, 2,200 years later, the diversion project is still serving the people of Sichuan, but the Min, unfortunately, has seen better times.

The reason is quite clear. The upper reaches are a combination of 138 rivers and streams spilling out of the mountains in Aba Prefecture. Before the Song Dynasty, the area was nothing but forest. Years of cutting, especially the past fifty, have brought the forest coverage down to a tiny 18 percent in the mountainous area.[28]

For a long time many people thought the upper reaches of the Min were a dense green carpet of thick undergrowth. In fact, quite the opposite is true. Apart from terraced plots on the steep hillsides, the area is almost nothing but barren hills. Below the thin, loosely packed soil there was nothing but ghostly white rock.[29]

As reservoirs on the upper reaches of the Min were gradually being destroyed, the region's agriculture placed greater demands on water projects such as the massive Dujiangyan system. In 1949 it diverted water to about 188,000 hectares of land for irrigation. Then the irrigated area grew to include the entire Chengdu Plain, and a large tunnel was even dug through the Longquan Mountains to extend it to central Sichuan. By 1993 it irrigated more than 670,000 hectares.

Unfortunately, this increase in the demand for water coincided with a drop in the water levels on the upper reaches of the Min. From the 1930s to the 1980s the flow dropped by 3.2 billion cubic meters, or nearly one-fifth.[30] And the gap between the flood levels and the regular level grew wider.[31]

By the summer of 1998, for the first time in history, the upper reaches of the Min had dry spots. When the rains came, water levels rose precipitously and went over the banks, causing serious mudslides. The dry valley simply could not absorb the rapid increase. The Min ran amuck and became as wild as the Yellow.

Meanwhile, the Zagunao, a tributary of the Min, developed a 100-kilometer stretch with an almost semitropical desert climate. In a semi-tropical climate, deforestation of a river valley can affect air temperatures and lead to increased evaporation, which increases desertification and the growth of desert plants such as cactus. That process is difficult to reverse because it is nearly impossible to bring vegetation back to arid and hot ground. The Min valley is said by a certain international organization to be "in danger of desertification."

But back to the Yalong and Dadu river valleys. Their annual water flow has gone from 14.8 billion cubic meters in the 1950s to 13 billion. Even more seriously, the glaciers that dot the upper reaches of the Yalong have shrunk 200 to 400 meters in the past few years, while the valley itself is beginning to show signs of desertification. The amount of riverbed with dry spots has grown from 200 to 400 meters.

Then there is Sichuan's Qingyi River, whose upper reaches have an abundance of rainfall. There are 379 square kilometers of primeval fir forests and 178 square kilometers of various broadleaf trees at middle and lower levels that traditionally provided stability during periods of runoff. But by the 1970s a large water diversion project was built on the Yuxi, on the upper reaches of the Qingyi, that could divert 700 million cubic meters of water annually. And over the past twenty years, state companies and collectives in the area have cut a large amount of timber, causing many mountain streams to dry up.

These changes have had a dramatic effect on the flood-drought cycle in Sichuan and other areas along the Yangtze's upper reaches.[32] In the 1950s serious droughts hit Sichuan about once every three years; by the 1960s it was about every two years; by the 1970s, eight out of every ten years; and by the 1980s they hit some areas at least once a year. This does not bode well for the future of the Yangtze.

This phenomenon can be seen elsewhere along the Yangtze's upper reaches. In Gansu province, the Bailong and Baishui rivers are the major source of water for the Jialing. The previous section spoke about the enormous deforestation in that area. That also caused significant changes in the local climate and an ecological imbalance that will most likely lead to an increase in droughts.[33] The changing drought conditions have had an adverse effect on the Jialing. The water reduction in the Yangtze has its roots in the remoter regions at the headwaters.

In the summer of 1978, after decades of looking, a team of explorers decided they had finally found the true source of the Yangtze when they entered an area between the Gelandandong and Tanggula mountain ranges on the Qinghai-Tibetan plateau. They discovered three separate streams: the Qumar (Chumaer), Dam Qu (Danqu), and Tuotuo.

The region was 4,500 meters above sea level. It had thin air, a cold climate, variable weather conditions, and plenty of ice. It was, in short, inhospitable. But it was also home to many species of wildlife, including yaks, wild donkeys, yellow goats, and the Tibetan antelope. Not long afterward it was revealed that the hide of a Tibetan antelope made into cashmere could bring as much as U.S.$30,000 on the international market. Hunters rushed in and killed not just thousands of the antelopes also but wolves, rabbits, lynx, marmots, and foxes — anything that moved, in fact.

In 1993 another group of explorers went to the region. They could hardly recognize it. In only a few years' time the entire population of Tibetan antelopes had practically disappeared, as had the wild donkeys, foxes, and wolves. Taking their place was an army of rats that could be seen scurrying along the roads, even in the daytime.

These rodents were no laughing matter. They burrowed into the ground and specialized in eating the roots of the grass that used to be so lush. When high winds swept the region, what was left of the grass was easily uprooted, leaving nothing to hold the earth when the heavier winds and snow came. Within very little time, what was once a vast playground for an amazing array of wild animals had become nothing but barren land.

One study in March 1999 showed that desert covered 19,058 square kilometers of the region and that serious soil erosion affected 106,000 square kilometers. The region had previously had swampy areas, but 90 percent of that swampy land near the headwaters had virtually dried up, as had many of the smaller streams.[34]

This all had a deleterious effect on the Yangtze at its source. When the 1998 flood hit lower down, everyone was being very gallant and cheerful and comradelike about the battle against the mighty floodwaters. The following year there wasn't much to cheer about when another crisis hit — a major drought. This takes us back up to the headwaters.

In March 1999, as a result of the declining amount of water on the upper reaches, ships plying the waters of the Jing near the city of Shashi found themselves running aground in some surprisingly shallow waters. By midmonth, the water level in the main channel at Shashi was breaking records in reverse: 2.27 meters deep. It hadn't been that shallow since 1903.

Similar problems had been seen on the Yangtze River delta the previous December. High tides entered the river near Shanghai seven times. Saline levels in Chenhang Reservoir — the city's second most important source of water — rose dramatically (above 500 parts per million) and for twelve straight days the city got no water from that source.

In the past two decades, the Yangtze's average annual flow has dropped 100 billion cubic meters. The official prediction is that a quarter of the Yangtze's water will be used by human beings annually by 2030. Taking into account the massive project to transfer water to the North China Plain, it is conceivable that the current torrent could come to a stop during the dry season someday. Then what?

RED DESERTS

The Xiang River is so clear that the pebbles lying fifteen meters down can be seen as clearly as pieces on a go chessboard.
    —*Notes of Hunan*, Luo Han (ca. A.D. 360)

AS WE HAVE SEEN, THE YANGTZE RIVER VALLEY sees 2.4 billion tons of soil erosion annually, with 1.8 billion tons coming from the upper reaches and 600 million tons coming from the middle and lower reaches, below the city of Yichang. It flows directly into Dongting and Poyang lakes, silting up the two largest catch basins on the Yangtze.

Dongting is on the northeast corner of Hunan province and was China's largest freshwater lake for centuries. Four major rivers—the Xiang, Zi, Yuan, and Li—drained almost the entire province before entering the immense lake. Poets and literati celebrated the beauty and clarity of these rivers.

These days, if one chances to fly across the region, it's almost impossible not to be startled by the uniformly brownish red color of the network of rivers meandering through the green paddy fields that jut up against the reddish clay hills, which are in turn punctuated by clumps of bamboo and stands of scrubby pine trees. That reddish clay soil forms a thin veneer over most of Hunan. It was the deforestation of parts of the province that caused the rivers to turn color.

For much of the last five decades of the twentieth century, China lurched crazily from one radical reform policy to another—complete collectivization, making worthless iron in backyard furnaces to the Antirightist campaign to learning from model oilfield workers and peasants to a rural family responsibility system—and Hunan, the home of Mao Zedong and other important leaders, and its people suffered one cruel blow after another. But the policies not only affected people's psyches, but affected the province's forest cover.

In the last twenty to thirty years, Hunan's forest reserves have shrunk 35 percent, by 100 million cubic meters. The soil erosion situation is no better: 47,000 square kilometers have lost an immense amount of topsoil, 6 billion tons in more than forty years, with an annual loss of 3 million tons of organic soils. Forest cover in the hilly regions has been almost completely destroyed, as has the ability to retain soil in denuded areas.

As the soil goes, so goes the plant life, especially on the hills. The result is that 37,000 square kilometers of good soil have become desert, something the locals call "red desert."

There is a stranger situation, though. It's just as someone here said: "If for three days no rains come, the river dries up, but if there is slight precipitation, there is a flood." That's how it is at the headwaters of the Xiang River. In 1998 the river flooded. That was followed immediately by a major drought. Thousands of streambeds dried up for a period of time, and numerous reservoirs and ponds dried up completely. Several hundred thousand people experienced shortages of drinking water.

Similar ecocatastrophes have hit Hunan's western mountains, where, during the Song Dynasty, the Chinese rhino still roamed the hills, as did the wonderful panda. Prior to the 1960s, tigers could still be spotted in the remoter parts of the province. In Tongdao county, near the headwaters of the Yuan, from 1954 to 1958 a local wildlife station ordered more than 1,000 of these animals killed. But even if those poor beasts had managed to survive for a few more years, they would have been left without a habitat.[35]

The destruction of the ecosystem is also evident in Xinhua county, on the middle reaches of the Zi River. The area has had soil erosion since the 1950s, but it has increased threefold in recent years. There was similar deforestation in Anhua county, also on the middle reaches of the Zi.[36] The loss of forest cover has caused soil erosion on nearly 30 percent of the Li River area and has made it Hunan's worst eroded area.

The sediment levels in these rivers have risen substantially. Recent measurements have shown certain parts of the Xiang, Zi, Yuan, and Li riverbeds to be up as much as 1 to 2 meters, causing some of them to resemble the Yellow, the "floating" or "hanging" river. There is sediment buildup in 30 percent of the province's 258 reservoirs, reducing their storage capacity by at least 550 million cubic meters in all.

Now, the Three Gorges Dam is supposed to be able to hold a lot of sediment and slow down the silting process that affects Dongting Lake. But at the same time, something needs to be done to save this major Yangtze River catch basin by stopping people from turning fertile land into red desert and by reducing sediment levels in the four rivers that feed Dongting.

The Yangtze flows east from Dongting to Poyang, the second of the "two lakes." The 3,000-square-kilometer Poyang has in recent years replaced Dongting as China's largest freshwater lake. The region around it is also a vast network of branches and tributaries of the Yangtze, such as the Xiu, Gan, Fu, Xin, and Rao, all of which spill into the great waterway in Jiangxi province.

The land here also consists mostly of red earth. Jiangxi used to be known as a particularly verdant area, but it too has been the site of an enormously destructive process, especially in its many hilly or mountainous areas.[37] Soil erosion has caused these rivers to turn the same reddish brown as those in Hunan.

Perhaps no place expresses the seriousness of the problem more than Xingguo county, which, it has been estimated, has undergone some of the worst soil erosion in the world. In a 669-square-kilometer area, average soil loss has been put at an astounding 13,500 tons per square kilometer. In 1980 China asked a prominent soil conservation expert who was a member of the British Royal Society to come have a look at the large eroded granite and purple rock formations covering mountainous parts of Xingguo. After he examined the scene, his only comment was: "Of all the serious cases of soil erosion I have witnessed on this planet, I have never seen anything like what lies before me. Nowhere have I seen such a serious and extensive case of soil erosion as here in Xingguo." Xingguo rapidly gained limited fame as another red desert.

These comments set off alarms in people's heads. There were efforts to try to solve the problem, at least in Xingguo. A massive reforestation campaign began in 1983, and the total area of soil erosion in the county dropped to 813 square kilometers by the early 1990s. But a little success in one county was hardly indicative of the situation as a whole in Jiangxi. In fact, by 1993, the province had a total of 54,500 square kilometers of eroded land, or 31 percent of all its arable land. That was four times the size of the eroded land recorded in the 1950s, and a 60 percent increase from the 1980s.

At present, Jiangxi is estimated to have 200 million tons of soil erosion annually. The beds of its major rivers have risen by 1.5 to 4 meters over the past fifty years. Meanwhile the lake bottom at Poyang is rising 2 to 3 millimeters annually.[38]

Solving the red desert problem will not be easy. Of the 2,180,000 square kilometers of red veneer on the face of southern China, as many as 800,000 square kilometers have serious soil erosion problems. This red earth supports nearly half of China's people. In view of that fact, not tackling the problem might be considered a crime. In any case, it would certainly be a terrible legacy for future generations.

IS IT UPSTREAM? OR DOWNSTREAM?

You live by the headwaters of the Yangtze,
And I live at its end.
We cannot get together my dear,
But we drink from the same stream.
        —Li Zhiyi (A.D. 1038–1117), Song Dynasty

THERE'S A SAYING IN CHINA: "The quality of the tea depends on the water with which it is brewed." In ancient times, the waters of the Yangtze's lower reaches were deemed by some Chinese tea connoisseurs to be the best of all.

Even today, when compared with the water of the Huai, Liao, or Yellow, the water of the Yangtze is still relatively clean. Numerous environmental reports have concluded with the statement "Overall water quality in the main stream of the Yangtze is still good."

Still, it's difficult to imagine a tea lover who happened to visit the upper reaches of the Yangtze taking a fancy to its waters and dipping a ladle in it to brew that really special cup of tea. At the Three Gorges, where the poet Du Fu once saw mountains of verdure filled with screeching gibbons, there is now nothing but barren, rocky hills and dried-up valleys, with nary a creature in sight.

The river is an even more shocking sight. Not only does the murkiness offend the eye, but the vast open sea of garbage and sewage is singularly striking. The river carries an endless stream of traffic, from tiny fishing boats to larger goods haulers to an ever-increasing number of tourist boats and container and ore ships. Bit by bit, all the Styrofoam lunch boxes and vegetable scraps, toilet waste, cooking oil, machine oil and industrial muck, and just about every other piece of garbage the river people can get their hands on is casually jettisoned over the sides of this mobile fleet of what appear on closer examination to be garbage scows.

In addition, over the past several years there have been a series of major accidents that caused boats to spill toxic chemicals and fertilizer into the river, leaving a sort of environmental time bomb.

But as unimaginably bad as the river traffic is, there is an even bigger contributor to this flow of filth. That is the cities that sprawl along the banks and creep up into the hills along the upper reaches and the seemingly countless number of towns and villages. Day by day, week by week, month by month, in a monotonously inexorable fashion, they throw or pour the detritus of 400 million people into the waterway.

A large amount of untreated sewage comes directly from the 394 outlets that begin at the city of Panzhihua on the upper reaches in Sichuan and extend downstream to twenty-one separate municipal areas. The large cities of Nanjing, Wuhan, Shanghai, Yueyang, Chongqing, and Zhenjiang all do their own little bit to add to the 500-kilometer garbage dump.

The city of Yibin is located where the Jinsha becomes the Yangtze. Just about every day of the year one of its leather factories releases untreated white and blue industrial waste into the river in a lethal carcinogenic cocktail at twenty times the acceptable level.

The Yangtze gets 6.3 billion tons of pollutants and waste from these cities, the highest amount for any river in the country, including the Yellow. But that is still only a fraction of the 20 billion tons of waste that comes from factories and cities and towns along its many branches and tributaries.

The Chengdu (Sichuan) section of the Min, its major feeder, is a bad example. Local people recall how during the 1950s not only were the levels of the Min quite high, even during the dry season, but more importantly the waters were very clear. Then water flow from the upper reaches began to drop, and there was a perceptible increase in the amount of untreated waste being dumped into the river. The water quality deteriorated.

By 1998 the situation was so bad that 132 National People's Congress delegates from Sichuan signed a petition demanding that the authorities clean the river up. The reason for this hardly spontaneous or timely action was clear. In April of that year a serious pollution spill on the Min had caused tons of fish to die, something that caused an outrage among the locals, although one may be left to wonder just why they were so surprised. The provincial government did not take the matter lightly and immediately began an investigation. It found that along the middle and upper reaches of the Min there were more than fifty major sources of pollution that released 130 million tons of waste into the river annually.

Then there was the Longxi, yet another tributary on the Yangtze's upper reaches. It provided water for 2.2 million people in the Chongqing area. There were seventeen paper mills, a chemical fertilizer factory, and a brewery in the area, and in spite of the river's crucial role they all dumped their wastewater into it. There were clouds of white foam billowing up 2 or 3 meters from the water's surface and a marked change in its hue, a lovely mix of blacks, reds, whites, and yellows in what locals now refer to as the "rainbow river."

At the confluence of the Wu and the Yangtze is Fuling (Sichuan), where garbage floats in from both rivers and collides with a jarringly odd effect. Perhaps the most bizarre phenomenon is the entire island of garbage sitting serenely in the middle of the stream. It has taken years to create this monument to man's indifference and destructive tendencies that is now a kilometer long and a meter high.

You may think that this is a local Mount Everest of river-born filth, but it is not. Just downstream from the Three Gorges, behind Gezhouba Dam, there is a 3-kilometer-long, 10-meter high mountain of garbage pressing against the steel railings of the dam, refusing to go away and just waiting to be flushed downstream.

Obviously, people throw their trash into the Yangtze and its every branch because they know they live elsewhere. For centuries the Yangtze's swiftly flowing waters carried away a certain amount of the sewage, although not all. But in the last two decades the amounts of sewage and wastewater have increased so dramatically that no river on earth could possibly carry them away. And just about everyone gets hurt, because in the end we are always downstream from someone else.

If the situation is not corrected, each and every major stream feeding into the Yangtze above the Three Gorges Dam could quickly turn the dam's 600-kilometer-long reservoir into the world's largest cesspool. It is hard to imagine a worse possible environmental nightmare than that—not impossible, but hard.

EAU DE VIE? OR POISON?

You irrigate our land with your clean stream . . .
You feed our nation with your sweet milk . . .
　　　　　—*Song of the Yangtze*, Hu Hongwei (1984)

WHILE PEOPLE LIVING ALONG THE LOWER REACHES COMPLAIN about the sewage coming from their brethren upstream, they in fact contribute more pollution that annoys people even farther downstream. Some of the Yangtze's heaviest polluters— Shanghai, Wuhan, and Nanjing—are on the middle and lower reaches and could be said to be the real vanguard in bringing high levels of pollution to the river.

When the river flows out of Sichuan province, the water is still somewhat clean, but it is hit immediately in Hubei by waste that continues to be discharged all across the province. Along that 1,046-kilometer stretch, there are several major cities that dump 2.1 billion tons of waste and sewage into the river in all.

Wuhan is on both the Yangtze and the Han. It suffers from a dearth of potable water for one simple reason: By the time the Han River reaches the city, its water is sometimes so filthy that it can't even be handled by the local purification plant. One major purification plant on the Yangtze had to be moved 300 meters out into the river so that it could avoid taking in the water close to shore, which was untreatable because of pollutants coming from a large paper mill.

The Han is the Yangtze's longest branch and has high levels of pollution. It flows past cities such as Shiyan (known as "China's Detroit"), Xiangfan, Jingmen, and Tianmen, which have a number of auto plants and chemical industries that have blithely used the Yangtze as a convenient garbage dump. In 1997, Hubei dumped 630 million tons of wastewater into the Han. The river's major branches all suffered the same fate.

The water pollution situation is no better on the vast Jianghan Plain, where the water problem has two sources: a drastic increase in the amount of wastewater and the systematic destruction of the province's many lakes.

There were more than 100 lakes in the countryside around Wuhan in the 1950s. Then the large land reclamation program got going, and by the 1980s that number had dropped to only 35.

Now, after two decades of the post-1978 economic reforms and no letup in land reclamation, the number has shrunk to 27.

The dumping of wastewater into Wuhan's few remaining lakes still goes on, day in and day out, as do the dead fish floating to the surface, a disturbingly commonplace occurrence in Wuhan. The popular local type of carp, which has been scooped out of local lakes and served up as a delicacy since the 1980s at Wuhan's best fish restaurants, no longer makes people lick their chops or feel very comfortable when they know that this particular fish can ingest toxic waste better than they can. But in the final contest between the fish and the ever-present toxic wastes, everyone in Wuhan knows who will come out the winner.

The same sorry state occurs just south of the provincial border in Hunan, where pollution levels are no better. And the Xiang River is a good example. As late as the 1970s, it still had clear water and plenty of fish just waiting to be transferred to the table. But as the chemical fertilizer industry gradually got hold of agriculture, the entire river area started going downhill. Farmers increased the use of artificial fertilizers, and the heavy runoff carried them into rivers and lakes, giving them some of the highest levels in the nation.

By 1981 that area was using 110,000 tons of chemical fertilizers annually — 150 kilograms per hectare — to produce rice, vegetables, and fruits with some of the highest levels of organic chlorine recorded. The use of fertilizers generally coincides with the rainy season, which begins in early spring, so it's easy to see why such large amounts of these chemicals end up in local streams, lakes, and rivers without ever being absorbed into the soil. They eventually end up in the Xiang.

It is may be too late to rescue some of the fish that were so plentiful in the Xiang back when there were 144 varieties swimming around in its waters. In the 1950s the catch in Hunan was 10 million kilograms. By the 1960s that had dropped to 8.8 million, and by the 1970s it was only 4 million kilograms. Although precise figures are not available, I have been told the situation was even worse in the 1980s, by which time certain highly prized fish were virtually extinct and the number of varieties had dwindled to forty.

Even the cities and towns in the hills and mountains of western Hunan, which is revered as one of China's most elaborate repositories of mythology, could not wall themselves off from modernity and escape the inexorable march of pollution, principally from the mining of nonferrous metals and the chemical industry.

There were dozens of poorly built factories on the upper reaches of the Huaheng, western Hunan's largest river, which dumped more than 3 million tons of toxic industrial waste into the Huaheng annually. They also produced 200,000 tons of mineral spoils and runoff from them. The mercury concentration in the river is 15 times the national standard; that of chromium is 4.6 times; zinc, 3.58 times; and manganese, 13 times. The river is a threat to all living creatures.[39]

We jump from there to the Lake Tai valley in southern Jiangsu province, which was known as a "land of fish and rice" — or it was until industrial plants began spewing forth poisons that caused the water quality of Lake Tai to deteriorate beyond the point of return. Some cities, such as Wuxi, have spent millions of yuan laying pipelines that could bring in the relatively clean waters of the Yangtze River, then they turned right around and laid other pipes to carry away untreated wastewater to the river to try to reduce pollution levels in the lake and improve the standard of living — but only in their immediate vicinity.

Farther downstream, as the Yangtze River gets closer to the sea in the delta region, it passes China's largest commercial center, Shanghai. The city does not lie alongside the main stream but is crossed by two tributaries, the Suzhou and Huangpu rivers. For years the people of Shanghai used these as the source of water for drinking and other purposes — but no more. Both have been horribly polluted since the 1960s, so much so that the water has turned distinctively black and exudes a stench constantly. It is completely unsuitable for human use.[40] Urbanites were left with no choice but to bear the huge cost of moving Shanghai's water purification plants farther upriver, where the water was still susceptible to treatment.

In the past, whenever water from the Huangpu became unusable, the city would draw on Lake Tai and the Yangtze itself. But now the lake's waters have gotten more polluted, so the only choice left is the Yangtze. Shanghai is now known as a city without a water supply. Shanghai has had to come to grips with the effects of other people's pollution dumped into the river upstream, but it adds to the problem also. It dumps massive amounts of wastewater into the Suzhou and Huangpu before they spill into the Yangtze. Those wily folks of Shanghai, ever business-minded, were able to secure a U.S. $150 million loan from the World Bank to lay a series of pipes to channel their wastewater away from the city and directly into the Yangtze. When this project is completed, 5 million tons of the city's wastewater will be dumped every day into the main stream of the Yangtze.

For generations, the waters of the Yangtze were compared to a mother's milk for their nurturing effect. Today, by the time the nurturing waters reach the sea they contain over 20 billion tons of various types of waste and poison. How long can this go on? Or rather, what does it mean for the future if nothing is done?

"THE ENDLESS YANGTZE"

Out of the depths of misfortune comes bliss.
    —*Book of Changes* (ca. 1000 B.C.)

THE READER IS NOW WELL AWARE of how the Yangtze was mismanaged for the past five centuries and, even more so, for the past five decades. One should be able to understand how a relatively small flood can cause a large disaster, like the one in 1998. The question remains: What is to be done with the Yangtze?

There is no end to the number of views and the answers to this question, ranging from the highly fanciful to the fairly practical. Some of the better ones come from scholars and specialists in water management and related fields.

One group argues that simply mending the dikes here and there produces a Band-Aid effect and cannot guarantee the safety and security of millions of people. They suggest spending several hundred billion yuan to consolidate and raise the embankments several meters to remove the flood threat once and for all.

That would also mean removing thousands of people from ancestral homes and taking large patches of fertile land. Even if this approach were affordable, it would not address the serious problem of sedimentation and would end up nowhere. Even now, the middle reaches of the Yangtze have, like parts of the Yellow River, become a "hanging waterway," in some places as much as

20 meters above the surrounding countryside. That leaves us with the question of just how much higher the banks can be raised.

Another group suggests diverting the river's waters into two ancient courses in Hubei so that floodwaters would be distributed equally between Hunan and Hubei. This book has already given a brief history of the problems that arose when these two neighbors were pitted against each other in the water diversion question—a conflict that has lasted for hundreds of years. Briefly put, this approach cannot be considered feasible.

A third group has come up with a more imaginative, daring proposal. Under this method, three major rivers—the Jinsha (Yangtze), Lancang (Mekong), and Nu (Salween)—could be exploited up near their headwaters at their confluence in Yunnan province by building diversion channels and underground waterways that would connect them. Then large dikes and retaining basins would be built and waters from one could be diverted to either of the others when levels were high or low.

This approach has its problems, too. First, the height of the mountains and depth of the valleys in that part of China are not exactly suited to such a large-scale construction scheme. Second, and more important, it would lead to more interprovincial and international conflicts.

Then there is the group of hydrologists who argue that we can simply dredge the elevated riverbed in the middle reaches and also dredge the estuary. Apparently this can just be adopted as an emergency method instead of a long-term strategy.

But a far larger number of people have pinned their hopes on the Three Gorges Dam and other large reservoirs. Leaving aside the obvious political, economic, and bureaucratic appeal of such a grandiose project, there is still the question of whether this and other huge reservoirs really solve the Yangtze's manifold problems.

The question was earlier posed about whether the Yangtze was fast becoming a second Yellow River. But the question about desertification in areas along the upper reaches was muted. It might be time, then, to return to the beginning and take a look at the controversy surrounding this matter. The following section zeroes in on the issue of forest cover and reservoir construction.

The people on one side of this issue argue that soil erosion in the Yangtze River valley is far more serious than has been thought and that the many reservoirs and water control projects built along the waterway over the past few decades are essentially worthless. The reason, they say, is simple: Erosion eats away 1.2 billion cubic meters of soil annually. That is enough to fill twelve large reservoirs every year. These people believe that the Three Gorges Dam is in no way unique and cannot therefore escape this fate.

From the other side we hear that the above figures simply show the enormous contribution made by reservoirs on the upper reaches. This argument concludes by saying that even more should be built and that this would be the ultimate solution. They do admit that the Three Gorges Dam will leave a very bad sediment problem, but the next step logically would be to build more dams upstream to stop the sand.

The opponents of reservoirs note that no matter how many of these things are built or how big they are, there is not one that can compare to the beneficent effects of forest cover, especially along the upper reaches. Fifty years ago, in western Sichuan and northeastern Yunnan near the river's headwaters, the primeval forest itself could take in or store as much as 400 billion cubic meters of water. But the systematic forest destruction of the past five decades reduced the forest's ability to drink up water to only 100 billion cubic meters. The gigantic Three Gorges Reservoir holds just under 40 billion cubic meters of water.

The advocates of large-scale projects such as the Three Gorges Dam like to note that reservoir construction and soil conservation are not mutually exclusive and that the two can go hand in hand. But, they say, the soil conservation work so far has been far from ideal and that the benefit of stands of young trees is not significant, which is the major reason why reservoirs fill up with sediment so quickly and why flood controls cannot depend on afforestation alone.

Environmentalists admit that the reforestation work has had less than stellar results, but they blame the lack of funding for that. And the lack of funds, they point out, is a direct result of the government having spent every last penny on huge dams.

The debate has gone on now for five decades, during which time more huge dams appeared along the Yangtze's drainage area, while green hills became barren. The Yangtze followed the route of the Yellow. The people living along the river were eyewitnesses to its diminished state, and many began to lose hope. The debates had apparently got them nowhere.

Chinese are fond of their pithy old sayings and love to fit the saying to the situation, or vice versa. One saying is "When misfortune reaches the limit, good fortune is close at-hand." For some, the flood disaster of 1998 was a godsend. The government's belated response to this crisis was to promulgate a policy of sealing off the mountains to protect them from further cutting, planting trees, giving up hillside farming and restoring the original forest cover, returning reclaimed land to freshwater lakes, giving up levees to release floodwaters in low-lying land, providing work as a form of relief, resettling people and building new towns, consolidating banks and dikes, and smoothing out riverbeds.

Well, this seemed to be an encouraging shift, since priority was obviously being given to restoring the ecology, while engineering projects were mentioned only as supplementary methods.

In the wake of the disaster, the government sent shock waves throughout the Yangtze River valley, bellowing, "Don't cut a single tree down!" It seemed obvious that, at long last, some of the top leaders had had their eyes opened to the real cause of this disaster.

Some Chinese possessed of vision and wisdom had been proposing this for the previous 300 years.

Two years before the flood, in 1996, then deputy premier Zhu Rongji paid a visit to the Dadu River in western Sichuan. When he looked at the endless kilometers of barren hills and watched tons of trees lashed together being floated downstream, he seemed a bit disturbed and commented: "You must be determined to put the lumberjacks in these mountains out of work! There need to be more trees planted and fewer chopped down."

And, after 300 years of ruthless destruction, there was finally someone to gather enough support for a restoration policy—but of course it took the 1998 flood to get someone to sound the alarm.

That many forest workers and lumbermen were the biggest fans of the policy shift comes as no surprise. After all, the forest cover had dwindled to almost nothing, and fourteen of the twenty-four major timber companies in western Sichuan were on the verge of bankruptcy. Nearly five of them had, by the 1980s, already been left with almost no trees to harvest.

Of course, they should already have stopped cutting at that time, but the government did not want to have to deal with the mess of dismissing or relocating them; it was just too much trouble. So for ten more years they kept extending the lumber roads at no small cost and cut down the few remnants of primary forest that were still standing in the almost inaccessibly high mountains. The workers suffered, the companies still ran in the red, and the best environmental barrier anywhere on the Yangtze was obliterated.

That leaves the question of whether the Chinese government can face dismissing these workers at this time. The biggest financial burden would be compensating the local governments and the people for destroying their "timber economy." The governments of that vast poor region depend on lumber sales for 70 percent of their revenues. Sichuan and Yunnan together would be denied 7 billion yuan in annual income from the once-vibrant lumber industry. And if this business were to go down completely, it would take various other related companies and industries with it, scattering unemployment all over the region. So it appears they're damned if they do and damned if they don't.

The next question, then, is the one about how the government can deal with this burden.

For centuries, peasants have waded into lakes to reclaim marshland to help produce their food and clothing. They contributed to the flood disaster in their own way, but they also became its victims. They simply chose to cling to their reclaimed land after so many disasters because it was the easiest thing to do and their living depended on it. So the government needs to find a way to help them escape this vicious cycle. And there's no easy answer here either.

A much more difficult but related question is how to get more than 18 million peasants to stop plowing the hillsides of the vast area that drains into the Yangtze. That would mean a loss of almost 10 million tons of grain annually. Just turning some of the gentler slopes into terraces, as some agroeconomists say should be done, would cost more than 20 billion yuan. Once again there's the question of where on earth the government would find the money for that compensation package.

Some of the skeptics living in the area adopted a wait-and-see attitude. Others just turned the night into day and cut quicker, which during the 1998 flood caused thousands of trees to pile up in the upper reaches of the Yangtze while the land along the lower reaches was being inundated by an immense flood.

But this time the government seems surprisingly serious. In late 1998 it began three projects: a natural forest protection plan, a returning-farmland-to-forests plan, and a returning-farmland-to-lakes plan. The goal of the first was to protect the remaining virgin forests. The government approved 96.2 billion yuan for forest protection, tree planting, and relocating several million lumberjacks who were suddenly out of a job in old logging areas.

For more than thirty-two years, the Sichuan General Forestry Bureau was largely in the business of making sure that as much timber as possible went down the Jinsha, Yangtze, and Yalong rivers. Now the organization has been formally renamed the Sichuan Yangtze River Forest Creation Bureau. Its 6,000 employees went from cutting trees to planting trees almost overnight. The compensation package has been costly—960 million yuan annually from the central and provincial government over the next thirteen years. Similar arrangements will have to be made in other provinces around the headwaters.

A second part of the project was meant to get farmers to give up farming and to start planting trees along the headwaters of the Yangtze, the Yellow, and other major rivers. The government set aside a large amount of money for both food and cash compensation and has promised to continue the program for five years.

The third part of the project was intended to get 2.5 million peasants to leave the lake areas along the middle reaches of the Yangtze. In late 1998 the government issued more than 10 billion yuan in treasury bonds to raise funds to move 1.5 million people from Hubei, Hunan, Jiangxi, and Anhui provinces and to build new towns to receive them. The project is still going on.

When one takes a closer look at these projects it is easy to see that they are moving forward with great difficulty. The major headache is still the shortage of funds. Behind that lies the angry struggle over resources by the ecological restoration proponents, on one side, and the people defending the vested interests of the dam builders, on the other. The latter camp is by no means coming out the loser in this fight, because a series of huge dams have been approved on the Yangtze and its tributaries as part of the government's grandiose new plan to open up the "western region" — meaning basically everything west of Beijing.

But we have seen an inchoate environmental protection policy, and there are those who believe it will grow. This section concludes with an optimistic note: "The Yangtze River keeps flowing eastward and our hopes and dreams last."

CHAPTER THREE

NORTHWESTERN CHINA

THE DANGER, MY FRIEND, IS BLOWIN' IN THE WIND

Where have you gone, oh beauty of Loulan?
   —Pop song from northwestern China

ONE DAY DURING MY EARLY SCHOOL YEARS, our adored teacher showed our class the most amazing thing: a topographical map of China. To the great surprise and delight of everyone in the classroom, here was the stark reality of China, laid out in a series of undulating lines of varying colors: 9.6 million square kilometers of territory. At least half of it must have been mountains, hills, and deserts. They just leaped out at you. Amazing. But then we saw that only 7 percent of it was arable land. There was a moment of concern among the pupils, but the teacher immediately calmed us down with the reassurance. "Look here," she said, pointing to the vast northwest, "this is Xinjiang, a vast area rich in raw materials and land and with very few people. Your task will be to assist in developing its nearly 200 million square *mu* [1 mu = 0.0667 hectare] of barren land."

What a relief! Here was a vast, open region that accounted for fully a sixth of China's landmass and had all the answers—clearly the country's salvation. What we didn't know in our childlike innocence—how could we have known?—was that this same vast territory was already having serious drought and desertification problems and that the massive reclamation work that had begun at midcentury had not left these 140,000 square kilometers—a territory larger than Alaska—untouched, quickly gobbling up both surface and subterranean water resources.

This vast area is one of the farthest places from any sea on earth, and it has the second largest sandy area on the earth's surface. The snowcapped mountain ranges that ring it are the sole source of water for its rivers and aquifers.[1] All but one of Xinjiang's rivers eventually seem to melt into thin air somewhere near the desert, but those points actually consist of wetlands, dense stands of shrubs like tamarisk, yantag, and calligonum. There are also many oases surrounded by Euphrates poplars. While it is true that many rivers that flow toward the desert rapidly evaporate, their banks are lined with green plant life, a fringe between it and the barren, parched soil of the desert. The rivers are the elixir of life since they support the ecosystem of a vast region.

But in the past few decades large tracts of barren land were converted into arable patches tailor-made for growing the thirsty cotton plant, and the region gradually lost its water resources. Today Xinjiang has 483 reservoirs with a total capacity of 8.5 billion cubic meters. That is larger than Xinjiang's annual average river flow, 7.7 billion cubic meters. But these projects helped the population grow tenfold, while oases increased threefold — and total lake area shrank to half its former size.

One of the lakes of Xinjiang that dried up was certainly the best known, Lop Nor in the southern part of the region. Until recently it had been China's second largest salty lake. When a team of scientists from the Chinese Academy of Sciences first visited it in 1958, Lop Nor was still a very active body of water, with birds all around it and schools of large fish, some of them nearly 5 feet long, the sight of which would leave anyone with a sense of awe at the vibrancy of life in the middle of a desert. But in just twenty years that lake was to become a vast sea of salty dust that put an end permanently to the special fish that swam in it and the wild animals that drank at its margins. More on the rapid drying up of Lop Nor later.

First we turn to the state of several major lakes in the northern part of the region. There was Lake Manas, long known for its fickle nature, which still covered a 550-square-kilometer area in the 1950s. To its east is the Gurbantünggüt Desert in the Dzungaria Basin, where the scenery was reported to have been magnificent. In the 1960s and 1970s, however, there was a large amount of reclamation work in and around Shihezi, an enclave of Han Chinese immigrants, directed by the Xinjiang Production and Construction Corps. A series of reservoirs was built on the upper reaches of the Manas River. The result? Manas Lake was deprived of its only source of replenishment and dried up in a flash, creating a new desert 1,500 kilometers long.

A similar fate befell Lake Ulungur, also in the northern part. It was also quite rich in the 1950s, with an estimated 870 million cubic meters of water flowing into it annually. Twenty years later, however, after hydrology and agricultural projects along the upper reaches of its major feeders had taken their toll, the amount of water flowing into the lake dropped precipitously. That killed off large amounts of vegetation around the lake and along riverbanks. The resulting desertification grew by 3,400 hectares.

To the west of these two lakes is Abi Nur, on the border with Kazakhstan. It has not dried up completely but has shrunk from 1,200 square kilometers to 500 square kilometers in the past fifty years. From 1949 to 1980 the population grew more than ninefold and farmland increased more than eightfold. Water consumption was up sevenfold. To support this increase, people dammed and diverted the lake's seven major rivers to irrigate cotton, wheat, corn, and fruits. And they built twenty dams on the Kuitun River alone, cutting off the largest tributary feeding the lake, which is now many times saltier than the sea. Large numbers of fish died, and birds fled it as if it were the plague. The bird paradise started turning into a dead lake.

In the 1950s there were fifty-two lakes in Xinjiang that were more than 5 square kilometers in size, with a total surface area of 9,700 square kilometers. But by the end of the 1970s that total area was a little over 4,000 square kilometers. A small group of lakes in the southern part of the Taklamakan and in the Gurbantünggüt deserts have just about disappeared.

The words of my primary-school teacher about the large amount of land in Xinjiang just ripe for the picking come back and are echoed still. Recent statistics show that a lot of land has been developed but that there are still at least 100 million mu yet to be cultivated. Many people, especially the officials, still think that Xinjiang is the most promising area for agriculture in all of China.

One has to question the reasoning of those decision makers and ask whether they really want to stay in the newly reclaimed land. Abi Nur is one example. A large part of the lake's bed is now exposed to the sun. And because it lies right in the middle of one of China's worst wind tunnels, which stretches over into

Kazakhstan, its salty earth has turned the farmland around its edges and even the Gobi Desert a grayish white.

The local ecology department has discovered that the wind blows 4.8 million tons of salty dust into the air every year. The dust is carried across the northern slope of the Tianshan range and Xinjiang's most developed areas and can reach Xinjiang's capital, Urumqi, 600 kilometers away.

But the local people are the primary victims. The white dust makes the grass wither, kills cotton plants, and has caused power blackouts by eroding power lines. It is especially severe in Jinghe county, where 70 percent of the grassland has been in a state of decline since 1980 because it lies downwind, where dirt accumulates. Lambs that feed on the local grass suffer from diarrhea.

Humans also suffer from health problems. A series of recent X-ray exams of 100 people in Jinghe showed 70 of them with lung diseases. Local hospitals have also reported an increase in the number of eye infections and skin diseases. Some people have voted with their feet; many retired officials who originally organized the reclamation effort have moved elsewhere. Sand dunes around Jinghe have been advancing by 30 meters a year, threatening 10,000 hectares of farmland and forcing many farmers to give up their homes.

The sand also threatens the transcontinental railway that links China's east coast with Central Asia, Russia, and ultimately Europe. More than 63 kilometers of track pass by the edge of the desert. Near Alar Pass in the 1970s, sand dunes were approaching the railway at a rate of 7 meters a year, which rose to 16 meters a year in the 1980s and 30 meters in the 1990s. In 1996 trains had to stop thirty times because of sand on the tracks.

Where water once shimmered in the sun, that same brutal sun heats the desert floor and sheets of sand cover everything. Xinjiang's deserts account for fully 60 percent of all of China's deserts. Desertification has gobbled up 800,000 square kilometers of land in the region, according to some estimates. And just how long can the islands of green last? Hard to say, but clearly the danger, my friend, is blowin' in the wind.

BAREBACK PONY

Poplar tree, poplar tree, you last 3,000 years.
You live a thousand years,
You stand another thousand years after you die,
And you lie on the ground without decaying for another thousand years.
——Uygur folk song

TARIM, IN THE TURKIC LANGUAGE OF THE UYGUR ETHNIC GROUP, means "bareback pony." Anyone with even a rudimentary knowledge of the river from which the Tarim Basin derives its name would understand where the name came from. It is in China's largest geological depression and traverses this vast area like a mustang pony, free from any human constraints, a truly wild river.

The Tarim River is 1,310 kilometers long, making it China's longest inland waterway and one of the world's longest.[2] Without the life-giving waters of this river system, it is clear that the Taklamakan would not have the many oases that bring relief to the low-lying Tarim Basin. *(Appendix 7 – Tarim River Basin)*

The very survival of 8 million people and much of Xinjiang's vegetation depends on this crucial waterway. For thousands of years the river gave Lop Nor its water and nurtured what is known as a "green corridor"—a 473-kilometer stretch of forest and vegetation flanking the old Silk Road, which runs along the Tarim's lower reaches between the Taklamakan and Kum Tagh deserts.

But catastrophe hit the Tarim River in the 1950s. The large reclamation projects along the upper reaches of its many tributaries added irrigation canals galore, substantially reducing the annual water flow in the main stream, with serious consequences.

Lop Nor soon dried up, but people seemed to take no notice at all. They continued building a large system of reservoirs that were sprinkled around the upper and middle reaches of the Tarim to serve the farmland expansion. This caused a drastic reduction in the amount of water flowing to the lower reaches. It went from 1.3 billion cubic meters annually in the 1960s to a mere 126 million cubic meters by the early 1990s.

But it didn't stop there. The reclamation work went crazy in recent years, with people living along the waterway digging recklessly into the sandy banks to divert the river water and squandering about 4 billion cubic meters in flood irrigation on the hot, dry desert.

Data from the Tarim River Bureau's Water Management Office indicate that at present there are 138 breach points along a 1,000-kilometer stretch of the river. Of the 50 largest, only 18 have an effective mechanism to control spilloff. For most, the diverted water never makes it to agricultural areas and is just wasted, an amount that could irrigate another 26,700 hectares of land.

One of the most destructive processes came about when large plots of land were planted in cotton, the most profligate, water-guzzling crop available—and, it so happened, the crop of first choice of the Chinese government and the locals, who were eager to reap the benefits of the post-1978 economic reforms.[3]

But an even more lethal blow to the Tarim came with the infamous Daxihaizi Reservoir on the lower reaches. The 180-million-cubic-meter reservoir could irrigate about 78,000 hectares of land, hardly a viable ecological trade-off. It is surrounded by desert and easily falls prey to the frequent sandstorms, which, it is estimated, can cause 200,000 to 300,000 cubic meters of valuable water to disappear in a flash. Even more disconcerting is the fact that the 4-meter-wide, 7,200-meter-long basic dam structure is nothing more than a pile of sand that could collapse at any time.

The tragedy of this worthless reservoir is that it robbed the Tarim of the last drop of water that should have flowed into the green corridor. The 340-kilometer stretch of the river's lower reaches dried up for twenty-seven of the thirty-seven years from 1960 to 1997, while from 1991 to 1997 only 6 million cubic meters of water was released into the lower reaches from the reservoir. In effect, the Tarim's lower reaches have dried up completely, with the result that there has been desertification around Lop Nor and its marshland and at Lake Taitema, which is now buried under 10 meters of sand.

Perhaps the greatest environmental impact of the drying up of this 340-kilometer stretch of the lower reaches has been on the once rich and vibrant green corridor, where dense forests of Euphrates poplar could counteract the parching effects of the desert. As the flow of water diminished in this eastern part of

Xinjiang, these forest reserves, like those in other parts of China, retreated and began disappearing in only a few short decades.

For thousands of years these poplar trees grew in China's northwest, living in complete harmony with the Tarim River. This was one of the world's finest forests, with trees rooted more than 10 meters deep in the earth, something of which the Chinese could have and should have been proud. But humans intervened and siphoned off the water from the main stream. Not only did the lower reaches of the Tarim dry up, but the water table sank from 2 meters below the surface in the 1960s to 16 meters today, completely robbing the trees of their sustenance. Even this most elegantly resilient and drought-resistant of trees could not hope to survive in such an environment.

Nor did the other forests and habitat at the Tarim's upper reaches fare any better. There agriculture intruded with an equally devastating effect. The Euphrates poplar would remain standing long after it had died, so they gave Xinjiang a ghostly, haunting museumlike effect with their different shapes and sizes. But the many wild animals of the region went as well. The Tarim tiger, which treaded the forests, and the large-headed fish that swam in the waters of Xinjiang have both apparently disappeared forever. As for the once-ubiquitous Bactrian camel, it is rarer than the giant panda of Sichuan. Unfortunately it lacked the cuddly cuteness to draw the attention of the world's cosmopolitan nature lovers. How can humans carry on in a place from which all these natural companions have been removed?

There are now about 8 million people living in the various oases of the Tarim River, spread over five prefectures, twenty-two counties, and in four areas with battalions of the Xinjiang Production and Construction Corps. Many are content with what they see as the achievements in this hostile land.

However, it came at a price. The natural oases were destroyed, while the desert swallowed up not just the green corridor but also thousands of hectares of grasslands. The immense Taklamakan Desert and the Kum Tagh Desert gradually merged.

According to the Xinjiang Ecology Institute, 94 percent of Argan prefecture, which is located along the lower reaches of the Tarim, suffers from serious desertification problems. In 1995, to try to save the green corridor, an emergency water release project was implemented. The effort was repeated several times, but it was too little, too late.

The dry patches are rapidly encroaching on the middle reaches of the Tarim, where for several years lakes have been drying up. In 1996 the middle reaches dried up for sixty days, a long enough time for the agriculture of the area to be affected.

In the past, the oases of the Tarim River valley would sometimes contract as a result of natural forces that would cause a reduction in the water flow throughout the valley. More recently, people have blamed global warming and other climatic changes. The fact is that since 1987, while global warming has gradually reduced the amount of snow in the Northern Hemisphere overall, in Xinjiang it has had the opposite effect. The increase in annual amounts of precipitation has had a positive effect in snow accumulation in the region over the past fifty years.[4]

According to Professor He Wenqin, a Xinjiang hydrologist, from the 1930s to the 1990s the amount of water flowing down from the mountains into tributaries and the upper reaches of the Tarim River increased. If Lop Nor had not been affected by human activity, any fluctuations in its water level would have been the result of natural causes, which, considering the increase in water flow, would have meant that the chance of a total dry-up would have been quite remote. Unfortunately, it was humans with a penchant for destructive behavior who caused it.

Anyone who visits the area of the lower Tarim these days can easily find deserted houses. They belong to units of the Xinjiang Construction and Production Corps, which arrived in the area in the 1960s with a "mission" to reclaim the vast amount of barren land for a glorious agricultural future. Forty years on, the region suffers from a dire water shortage on 40,000 hectares of land. Agricultural output has plummeted, leaving thousands of workers with little or nothing to do. Upward of two thousand have run away.

Gazing at these other ghostly relics with their empty windows, the legacy of the corps, it's difficult not to think of the ancient and long-lost city of Loulan.

DIVINE RETRIBUTION

The lonesome whirlwind forms a pillar between desert and sky,
The large setting sun glows above the long waterway.
     —Wang Wei (A.D. 701–61), writing about the Hei River

IT WAS THE SPRING OF 1988, and an unusual cold air mass hit Beijing, bringing with it something even more unusual—rainfall at a time of year when it was not so likely. After the initial shock wore off, the people of Beijing ran out and held out their hands to feel the drops of rain. To their surprise, what did they find falling from the sky? Mud, not rain. Over the course of the next few days, things turned stranger still. It was not only Beijing that had been pelted by drops of mud, but the area south of the Yangtze River. This preceded a sandstorm that had come from northern China and swirled through Beijing, then descended on regions farther south without losing an ounce of its energy.

Scientists wanted a quick answer, so they rounded up the usual suspects and quickly determined that the source of the mess was Inner Mongolia, especially Etsina, on the border with Xinjiang. But people were puzzled by how this region, long known for its oases, could have caused such a large quantity of sand to pile up in the capital.

Etsina is in the northwestern part of the Alashan Plateau in the western part of Inner Mongolia. It had a long period of prosperity mostly because of the lakes in its low-lying parts. The lake area was formed in a delta of the Hei River and was the only green spot between the desert in Xinjiang and that in Mongolia.

The Etsina lakes got their water from a river that was fed by the Qilian Mountains of Gansu province, to the south. The Hei was once so full that it could traverse a desert like the Badain Jaran and keep enough water to empty into the lakes. But it had the same sad history as the rivers of Xinjiang. The same irrigation campaign that started there was applied to the upper reaches of the Etsina. It drained away the source of life.

The Hei River used to empty into two lakes, Gaxun Nur and Sogo Nur, which were 30 kilometers apart. The waters of the two lakes mingled during the rainy season. Some photographs from 1958 show Gaxun covering a 260-square-kilometer area. It was a major summer habitat of swans, and it and the oases around it supported a wide range of other animals and birds such as black cranes, wild donkeys, sheep, rabbits, and lynx.

By 1961, however, Gaxun Nur had dried up, and a year later the same fate befell Sogo Nur. After more than a thousand years of glory, the Etsina lakes region had come to an end. The surface water disappeared, but gradually so did the subterranean water. On 50,000 hectares of land, two-thirds of the hardy poplars died of thirst, and of a million hectares of suosuo (*Haloxylon ammodendron*), there were only 200,000 hectares remaining.

Etsina's neighbors to the south on the upper reaches thought they had a good reason to cut the river off. They lived in what was known as the Hexi Corridor in the northern part of Gansu, which had long been thought of as the "frontier's granary." It had been part of the age that took grain output as the greatest sign of development, and the Hexi farmers had an edge over their herdsmen neighbors in Etsina because they had dibs on the glacial runoff from the Qilian Mountains.

The Hei was the largest waterway from the Qilian, and it naturally became the focus of water retention works by the Hexi people. From the 1960s on, there were dozens of reservoirs and dams built along it. In effect they blocked its flow, and the water that had normally made its way into the Etsina began to disappear along the corridor—so much so that even when floods hit the Hei River not much water made its way into the Etsina. And as the Hei's water flow slowly ground to a halt, it and the Etsina began to suffer not just seasonal stoppages but semipermanent dry spells.[5]

The Hei was not the only water system used by farmers. The same devastating effect was evident on the Minqin Basin in Gansu province at the end of the Shiyang, another inland river. As late as the 1950s, the Shiyang's waters were so plentiful that they had formed two 10-meter-deep pools on the eastern and western sides of the basin, which were havens for swans and other waterfowl and sustained a water table that was only 0.4 meter below the surface as part of a classic oasis ecological system.

However, in 1958 work on the Hongyashan Reservoir on the Shiyang's upper reaches was completed and the river water could be used for agriculture. It took some time, but the Shiyang's flow went from 578 million cubic meters in the 1950s to just 148 million cubic meters in the 1990s, and the related development made even the underground waters of the Minqin Basin dry up. Of the area's more than 26,000 hectares of arable land, nearly 20,000 have been abandoned because of irrigation difficulties, and they are on the verge of becoming deserts. Large tracts of local forests on more than 72,000 hectares of land have withered or died, and more than 67 percent of the area is desert.

The people of Hexi continued to divert water but they did not particularly show much humility toward the Qilian Mountains, whose precious water they were sucking up. They cut down trees on the mountainsides relentlessly for personal use and for industry and agriculture. Then there was the gold mining and other mineral extraction carried out on such a large scale that the mountains were devoid of natural resources, both above and below ground. The northern slopes are now barren, without even a trace of grass and certainly no meadows.

This degradation has caused the snow line and the glaciers to retreat in recent years, so the amount of water running down into the valleys has gradually declined. The effect was felt on the region's major waterway, the Shiyang. In the 1950s it had a flow of 600 million cubic meters, but by the 1990s that had dropped to less than 200 million cubic meters. There are eight major waterways with headwaters in the eastern part of the Qilian Mountains. Their volume dropped from 1.89 billion cubic meters in the 1950s to 1.23 billion cubic meters by the mid-1970s.

When the people of the Hexi Corridor, who were used to their bountiful water supplies, realized the seriousness of the situation, panic set in. The people set themselves to keeping as much water as they possibly could from all the major waterways that originated in the Qilian. They stole almost all the Hei's water, arguing that the water did more good in Hexi than it could in the swamps of Etsina.

The people of Etsina were, not surprisingly, not swayed by this argument and protested that the water was everything to them. Their voices were not heard. And in Etsina, as in so many other parts of China, the deserts began to advance at an alarming rate, eating up 3.3 million hectares of grassland and taking with them a large number of rare animal species. More than 20,000 people became what could be called ecorefugees.

But there was a kind of divine retribution, if one wants to call it that. It was determined and unerring. If Etsina could produce sandstorms that could reach Beijing or even the Yangtze River delta, then what chance could the Hexi Corridor possibly stand of avoiding them? It was hit by savage sandstorms with increasing frequency — in fact, more in the past fifty years than in the fifteen centuries from the Han to the Song Dynasty.

And May 5, 1993, was a day no one there will soon forget. The region was hit by such a wild sandstorm from Etsina that dirt and grit filled the air and wreaked havoc on cities and villages alike.[6] The storm buried nearly 100 people alive, many of them schoolchildren, throwing the people of Hexi into a state of despair. It is sad, but the wind and the sand have drowned out the sounds of the death throes of the last swans that decorated the waters of Etsina a long time ago. Is this divine retribution?

"GO WEST, BIG BROTHER, GO WEST"

In the immenseness of the clear sky and vast land,
The herds can be seen only when tall grasses dip low in the wind.
　　　　—*Song of Chile* (A.D. 546), by Hulu Jin, a nomadic tribe general

WE MOVE EASTWARD TO GANSU, then farther still to the Inner Mongolian grasslands, a 2,000-kilometer expanse north of the Great Wall. The area's climate is dry, with limited rainfall, rapid evaporation, and scant water resources. The rivers of Inner Mongolia are largely seasonal, with most water appearing in the summer and considerably less or none at all during the other seasons.

Nonetheless, the region's rivers and lakes used to be sufficient to support a vibrant ecosystem and the needs of a society consisting mostly of nomads who relied on animal husbandry for their livelihood. For centuries, the herdsmen roamed freely with their sheep, cattle, and horses across the land in pursuit of what little water was available and with little or no impact on the local ecological balance.

The reclamation and grassland conversion projects were not so kind. Reclamation was often followed by desertification, which led to reclamation of even more grassland, which led once again to more abandoned land.

For the lengthy period of time that this policy was pursued, the Chinese, most of them refugees of environmental disasters within the Great Wall, sadly humming the tune of a popular folk song, "Go west, big brother, go west," managed to turn what had been a vast green grassy land into an expanse of barren territory.

The population problem, a cause of all this, just never went away. The amount of available arable land shrunk while the number of people grew. They needed food, water, and everything else and were left with no choice but to pick up their belongings and go seek a better life elsewhere.[7]

This impulse became most ferocious on the Inner Mongolian grasslands in the frenzied years of the Great Leap Forward, when a major part of China suffered irreparable psychological and ecological damage thanks to the irrational Maoist policy of "planting grain everywhere." For centuries the nomadic herdsmen of Inner Mongolia had lived in a kind of balance with nature until they were pressured and cajoled by the cadres to settle down and switch to growing grain, which was incredibly ill-suited to this part of the country. A mere three years or so of crop cultivation on the exceedingly thin soil was enough to damage it.

By the beginning of the 1960s, in spite of putative attempts to retrench after the insanities of the Great Leap Forward, the Chinese government actually increased its production demands from the grassland from 50 kilograms per person to as much as 250 kilograms. Even as late as the 1970s, the amount of land inappropriately converted to grain and similar forms of production was five times the figure of the 1950s in the Wulan Chabu League near the Great Wall.

No matter how hard mankind might try to transform nature — perhaps the most serious flaw in the Chinese Communist mentality — the large pieces of arable land were often taken over more quickly by the forces of nature: the wind and the sand, which turned them into sand dunes, neither arable nor edible like grassland.

TABLE 1 Shift in Agricultural Yield in Huade County
of the Wulan Chabu (Ukan Ch'ap) League
(administrative area)

Period	Arable land (hectares)	Grain yield (kg per hectare)	Rainfall (millimeters)
1950s	32,000	535	396.7
1960s	80,000	277	319.4
1980s	120,000	272	294.1

From this one can see that thirty years of hard work produced a 375 percent increase in arable land, while per hectare grain production amounted to only 17 percent of the earlier figure and half of the original per capita yield. Rainfall in the area was only 74 percent of the figure thirty years earlier. In fact, by 1994, of the 5.53 million hectares of arable land in the various leagues of Inner Mongolia, more than 90 percent had been adversely affected by soil and wind erosion.

Barren land was taken for agriculture in the Wulan Chabu League, as was the case in many other parts of Inner Mongolia, and subjected to the same erosion process. During the Cultural Revolution (1964–76), various units joined the Production and Construction Corps in reclaiming grassland, 216,000 hectares in all. More than 3 million hectares of grassland was taken for vegetable gardens and livestock, and Inner Mongolia lost a tenth of its grassland. In the 1960s Inner Mongolia still had over 8 million hectares of grassland, but thirty years later that was down to just over 3 million hectares.

Inner Mongolia's grasslands were the largest and best pasturelands in China. But in the past five decades, not only did the grasslands become smaller, they became worse. Many herdsmen can look back to the 1960s, when their sheep ate only the tender tips of the grass. By the 1970s, when the grass was sparse, the sheep had only some stalks to eat. In the 1980s, they had to eat whatever grass they could find, and by the 1990s they had run out of grass and often had to depend on some kind of straw.

In recent years the grassland has been hit periodically by "white storms" in winter, when the snow killed off what little grass was left, and "black storms" in spring, when the snow evaporated into thin air and left a barren earth. Experts attribute these to the overgrazing and farming that leaves the patches of grass so sparse that whatever snow makes it to the ground is usually just blown back up into the air and does not last until spring to melt. Droughts follow, the grassland produces even less grass, and there is nothing to fill in the empty patches.

In more recent years, as China gradually turned itself into a consumer society, a great "wild herb" fad swept the grasslands, fueled by the appetites of the health-conscious members of the new society. In northern China there used to be more than 6,000 varieties of wild plants, including licorice root, truffles, the desert cistanche, the hair plant (*facai*), Chinese ephedra, Chinese caterpillar fungus, and a wide assortment of exotic other plants. In their natural state these plants had a dual purpose — they were hardy, close to the ground, and withstood the harsh winds, and they were a valuable source of nutrients.

Unfortunately, the humans were all too aware of this. From 1993 to 1996 some 1.9 million farmers from Ningxia and Gansu went in search of the valuable wild hair plant on a 1.5-million-hectare stretch of land in Inner Mongolia. It had become the latest fancy of the nouveau riche cognoscenti of southeastern coastal areas, not because it was such a delicacy per se but because its name sounds like the word for "fortune," especially in the Chinese dialect of that area. Unfortunately, as a result of this fortune, 1.3 million hectares of land were severely damaged and 400,000 hectares were seriously eroded. The gang of foragers was as bad as an army of ants, and they moved on from Inner Mongolia to Ningxia and Gansu in pursuit of licorice root.

The people had let the cat out of the bag, and now it has run amuck. Droughts and sandstorms have gradually become the region's routine. The droughts that have ravaged the region since 1999 have caused 31 million hectares of grass to wither and have caused half a million head of livestock to die. In addition, the region's topsoil is so dry that 1.3 million hectares of farmland have had to be abandoned.

To make matters worse, the region has also suffered from blizzards earlier in the year, followed by swarms of locusts, then sandstorms. The blizzards froze 300,000 head of livestock to death.

The disasters have affected 9 million people, and 3 million of them have been left in absolute poverty. Officially speaking, that means an annual per capita income of less than 625 yuan, barely enough to feed a person. The region had what seemed to be a fairly successful poverty relief campaign from 1994 to 2000 that helped 2.7 million herdsmen and farmers get out of poverty by expanding farming and increasing livestock, but that now seems to have come to naught.

This had not just an ecological impact but a serious meteorological and climatic effect as well. A large amount of sand in Inner Mongolia gradually shifted southward, causing greater aridity in the northeastern part of China. This caused more than a million tons of sand and dust to be blown into Beijing. So what? Well, Beijing curiously is now designated as a desert region, with a major concentration of sand and with dune formations only 70 kilometers from the center of town.

Then Beijing was awarded the position of host of the 2008 Olympic Games amid a great deal of ballyhoo, and the local government began pouring billions of dollars into tree and grass planting to show that the capital was up to snuff. Other cities on the North China Plain followed suit. But, leaving aside the question of just where the water for all these green projects is coming from, if we let the formerly rich and verdant green grass of Inner Mongolia disappear just like that, then all we can look forward to is more sand and dust in the air of Beijing, no matter how much green we try to lay down.

"Toward a Union of Humanity and Nature"

Human beings follow the Earth. Earth follows Heaven. Heaven follows the Tao. The Tao follows the way things are.
    —*Lao Zi* (580–500 B.C.)

For the past decade the Chinese government has committed substantial amounts of manpower and materiel to the western parts of the country. Now, after two decades of reforms and opening up, the state has launched yet another "go west" campaign that gives the north-central and western parts of the country priority in economic development.

Before the country goes all out in this effort, however, it would be helpful to bear in mind that in the country's western and northwestern parts there is a shocking scarcity of that most essential resource — water. This depressing fact, unfortunately or perhaps understandably, has been largely ignored by Chinese policy makers at the center (Beijing) and in the provinces to the west.

In Xinjiang the government's strategy dealt solely with the "black and white" — oil and cotton. In Gansu the slogan was "Create another granary for the frontier similar to the Hexi Corridor." In Ningxia regional leaders also worked hard to increase the amount of land under irrigation. In Shaanxi an enormous amount of work went into looking for coal, while in Inner Mongolia there was a search for new sources of water underground to support a much-hoped-for economic takeoff. Virtually all these places have plans to reclaim the grasslands and other barren areas for future development, but little or no mention is made of the cost or where the funds are coming from.

But let's back up a minute. Perhaps these dedicated optimists should take a closer look at some other places with similar development efforts that went horribly wrong. One dramatic example was the Soviet Union's Central Asian strategy in the 1960s for the area around the Aral Sea. For several years around 750,000 hectares of land were irrigated with vast quantities of water so that it could produce cotton. That meant severing both the Amu Darya and the Syr Darya, which flow through Turkmenistan and Kazakhstan, respectively. This did in fact support the people who moved into the area, and it did help create more wealth, but it was impossible to sustain. In thirty years, the incredible Aral Sea, once the world's second-largest salty lake, shrank from 6,700 square kilometers to 3,400 square kilometers, and its salt content trebled. This made twenty-four varieties of fish disappear and caused 60,000 fishermen to lose their jobs. That wasn't the worst of it. The lives of the locals were threatened by the accumulation of salt and other inorganic poisons in low-lying areas around the lake. Then there was the residue of fertilizers that had been applied heavily to the cotton crops. When the heavy winds, which were unchecked by any trees, swept the ground, all these chemicals became airborne. The incidence of respiratory diseases rose thirtyfold, and the incidence of water-borne diseases such as typhoid and hepatitis increased sevenfold. Prenatal death rates in the region went from 2.5 per thousand in the 1960s to a whopping 100 per thousand in recent years.

Unfortunately, the same perilous developmental road has been followed in western and northwestern parts of China, especially in Xinjiang and Gansu, where the water and soil conditions are not all that different from those of Central Asia. There is plenty of sunshine and sandy soil, which make it suitable for growing cotton, and the desert land can be transformed into high-yield cotton fields—as long as there's enough water.

In Xinjiang, leaders followed the same old "black and white" strategy, putting all their developmental eggs in one basket—oil and cotton, primarily the latter. And thanks to favorable tax policies, Xinjiang's cotton output grew rapidly even though there was a glut nationwide. In the year 2000 Xinjiang had a million hectares planted in cotton, with total output of 1.5 million tons, accounting for a third of the nation's total and 8 percent of the world's output.

In recent years in the Hexi Corridor there was an enormous increase in the amount of land devoted to cotton—30,000 hectares. Even people in places as unlikely as Etsina looked to cotton as the royal road to riches. And in spite of a substantial decrease in the water flow in the Etsina River, total cotton output set a record—2.4 million kilograms. This was cause for celebration, and the people of Etsina and those of Xinjiang and Gansu believed that their "silver triangle" was indeed the source of new riches. They paid little heed to the number of poplar trees that withered as the water resources were redirected to the cotton fields.

Lakes and rivers in these regions lost water rapidly, and people had to turn to deeper wells. They were aided by modern technology and sophisticated equipment that seemed so much more efficient in finding water belowground than the camels their ancestors relied upon. In recent years underground water exploration has been given priority, and the explorers were in the vanguard of the "overcome the desert" campaign. But the aquifers in the desert have limited supplies, so the search for more water has come at great cost to the environment.

One of the many examples of overdrawing on the dwindling water supplies was the region around the ancient city of Dunhuang on the Silk Road. That is the site of China's most incredible collection of Buddhist grottoes, which were a repository of important scriptures and are now a popular tourist site. The Crescent Spring there was famous for never having dried up in spite of being surrounded by desert. For hundreds of years it was the object of much mystery and lore.

But in the government's recent "go west" campaign, the area around Dunhuang has been developing at a rapid pace. The city even came up with the idea of a large development zone for industrial growth. That and other projects tapped into the aquifer, and naturally the Crescent Spring began getting less water. One interesting result of its rapid disappearance was the solving of the thousand-year-old mystery, as a study revealed that the spring was fed by a small underground lake that rapidly lost its water as well. Obviously that was a fairly backhanded way to go about solving the ancient mystery — it is hard to have a mystery without a mysterious source.

So the water resources in that part of China have become scarce and the ecosystem is on the verge of collapse. And the reason the situation is so bad is that man's ability to track down and divert water has improved substantially, combined with the large-scale industrial development of the past few decades.

The economic growth of the region came at the cost of the environment because it tried to make the region support more people than it could, as was the case in the Aral Sea project. Recent data show Xinjiang's population was about 2 million at the beginning of the twentieth century. A hundred years later it was more than 20 million.

In Zhangye prefecture in the Hexi Corridor the population in 1949 was 549,000, but by 1991 it had grown to 1.15 million. And in Inner Mongolia in 1949 the population was slightly over 6 million, but by 1998 that had jumped to more than 23 million. The population explosion has had the most deleterious effect on the struggle between humans and nature in that arid region.

But nature imposed its own restraints on development, in this case the limited amounts of water. This has become so serious that in spite of all the putative gains in productivity and material wealth that have come from abusing nature's resources, the losses will outstrip them several times over.

It seems a bit late, but one can still ask how this abuse of precious water resources can be stopped. Even as this is being written, water resources are still being abused in Xinjiang, Gansu, Inner Mongolia, and other parts of the west as a result of the same old inefficient and uncontrolled flood irrigation methods. In many places as much as 12,000 cubic meters of water per hectare are being used for agriculture when in fact 4,500 cubic meters would suffice. This is clearly an ecological crime. Its victims are the plants, birds, and animals.

But this is not at all surprising. In spite of its scarcity, the region's water is, like all natural resources and raw materials in China, still incredibly underpriced, just a part of China's much-vaunted "cheap water policy." And until the basic price of water is raised substantially, the waste, especially in agriculture and industry, will continue. What needs to be factored in, but in fact never was, is the ecological cost. We need to remember that for every extra square meter of cotton that is watered along the upper reaches of many of the rivers in the region, a poplar tree could bite the dust in the lower reaches. That is the simple exchange rate at the most basic level.

Those people who are opposed to tampering with these time-tested methods and the future development plans say, with a perverse bit of extremist logic, that "following your line of reasoning means there's no need to make any effort to develop the northwest at all. You're just saying that the people there deserve to live in poverty forever." It is certainly true that a substantial increase in the price of water would, in the short run, be a bitter pill to swallow, and it would certainly impose some limits on immediate development. But there seems to be little alternative if we want to leave a little water for nature and rescue the fragile area from irreversible destruction.

And not everyone is blind. When the degradation got so bad that it threatened the survival of human beings, some people started to make concessions. The Wulan Chabu League in Inner Mongolia, after years of ecological destruction, decided to look at the now-fashionable concept of "sustainable development." By 1994 the league had taken 530,000 hectares, a third of its arable land, out of agricultural production and turned it back into grasslands and forest. Thanks to a somewhat improved ecology, grain output in the league actually increased.

This encouraged other parts of Inner Mongolia, where they decided to return 2.7 million hectares of farmland back to grasslands by 2010. This painful transition needs help from prosperous cities such as Beijing and Tianjin, especially if they are really as tired as they say they are of the sand filling their air and their noses.

In fact, this project to return farmland to forest and grassland has reached the national policy level, and slogans about "restoring the green mountains and clear rivers in northwestern China" can be heard issuing from the mouths of the top leaders. There is still a deep fear, however, of the dark truth behind the great "go west" campaign, which is that the region will get more immigrants, more farmland development, more factories, and of course more ecological and environmental damage.

As the sands continue their inexorable march across northern China, the citizens need to choose between the quick fix of the short-term gains and a slower but more manageable process. For decades man has overstepped the bounds in exploring nature, and no one's face will be lost if a cease-fire is declared and concessions made.

CHAPTER FOUR

NORTHEAST CHINA

ONCE EVERY TEN THOUSAND YEARS

The provinces of Jinlin and Heilongjiang are full of primeval forests. . . . Local tribes are either nomads or hunters and they seldom cut trees. The forests there extend several hundred miles without roads and therefore provide better defense than the Great Wall.
—*Northern Defense Notes* (1858), He Qiutao, a Qing Dynasty inspector

IT WAS MID-JUNE 1998, during a trip to the northeastern province of Heilongjiang, on the border with Russia. I noticed the headline in a local newspaper: "The Songhua River Is Thinning Out." The word *thinning* actually concealed all manner of agony and pitiful conditions among 50 million people. But just as the media were giving a lot of coverage to the record low water levels in the Songhua, there were dramatic increases taking place in its upper reaches, where it is known as the Nen (Nonni).[1]

By the end of July, the flooding on the upper and middle reaches of the Nen had become a disaster for the people of the area, who were already suffering from poverty and all sorts of deprivation. The floods hit the larger cities and almost covered the critical Daqing oilfield, in a low-lying wetland. In the last-ditch struggle to protect the derricks and drilling sites, the floodwaters were sent into the surrounding countryside when some river embankments were blown up. The oil survived. The crops didn't.

Hydrologists reported record high water levels on the Nen's tributaries. On the Nao'er they were at flood levels that occurred once every 200 years; on the Nuomin they were the kind that came once every 500 years. On the Yarlung it was a 1,000-year flood, and on the Alun, incredibly, it was a 2,000-year flood. Together they brought an immense amount of floodwater tumbling down into the Nen, resulting in a 10,000-year whopper. *(Appendix 8 — northeastern rivers)*

The old folks said they'd never seen anything like it. But right up until the flood hit they were still humming the same old folk tune about the Songhua having "copper embankments and an iron bed" that could withstand any flood. This incredible concatenation of events had hydrologists shaking their heads and going back to the books to rethink the issue from an ecological angle — for the first time.

When the entire northern region of 1.42 million square kilometers was given a thorough "health check" by these scientists, they found that the northeast had suffered widespread damage similar to that of most other areas. But there was one distinct difference: The other regions had been experiencing this kind of destruction for centuries, whereas in the northeast it came much later, but with much greater ferocity.

The northeast was simply a paradise of trees and vegetation. Before it had been settled by people from China proper, down below the Great Wall, its forests had covered more than 90 percent of the territory. The expansion into the northeast was still very limited at a time when forests had already disappeared in many other parts of China.

As we saw in the Yangtze section, the Manchu Kang Xi Emperor oversaw what was perhaps the most frenetic period of land reclamation. But it was this same emperor who closed off the entire northeast in 1668. All logging, farming, fishing, and mining were strictly forbidden in the region. Any disobedient person could be beheaded. The northeast was home to the Manchu, and the superstitious ruler did not want any immigrants from down south ruining the region's auspicious feng shui.

That left northeastern China as one of the largest nature reserves in the world. According to the notes of Qing Dynasty inspectors, the vast primeval forests of the mountains and areas lower down were so thick that humans could not penetrate them. The rivers rushing out of the mountains were calm and clear. They left huge marshlands, and these and the forests proved to be an ideal habitat for tigers, bears, leopards, wolves, wild boar, cranes, and a wide range of other birds and beasts.

These restrictions held until the early twentieth century, but when the legal barriers to immigration were finally lifted, the great reserves of mature growth began being chopped down for land reclamation and agriculture.[2]

This was a shocking change. In the nineteenth century, the forest cover was still around 70 percent. But by the time the Japanese invaded the mainland, they were looking to Manchuria as an answer to their timber problems, and by the time the People's Republic was founded in 1949, less than half of the territory, somewhere around 45 percent, was still covered with forests.

Even so, the remaining forests were still vibrant primeval forests that were a vital part of the ecosystem, albeit one in much need of protection. Unfortunately, under the new government the whole process just picked up almost as if it hadn't missed a beat. And the people had no interest in the old maxims and saw no need to get all worked up about the feng shui of a defunct empire.

A WAKE-UP CALL FOR LUMBERJACKS

I'm an ambitious young man from the Tiger Valley,
Carrying an axe in my hand,
Chopping down trees on high mountains,
Giving my precious years to the state.

I'm an educated young man from the Tiger Valley,
A full-time lumberjack in my hilly town,
Trained with the best logging techniques,
This happy life will last 10,000 years.

—"Song of Lumberjacks," a great hit in the northeast in the 1960s

ACCORDING TO CHINA'S FOURTH NATIONAL STUDY of its forest reserves, forest cover in the northeast was as follows: 26.89 percent in Liaoning province, 33.6 percent in Jilin province, and 35.55 percent in Heilongjiang province. Now, while these figures might at first glance appear to inspire a certain amount of hope, a closer look reveals that exactly the opposite is true.

Half a century ago most of the region's forests were rich and dense. Today they are primarily second-growth timber put there by man. And while these trees grow rather rapidly, so that forest cover in Heilongjiang has actually increased to 41 percent, the board footage and the quality of trees have decreased substantially.[3]

The northeast's forests still provide a third of the entire country's lumber. But in the 1960s and 1970s it was one-half, and most of that timber still came from primeval forests. The biggest loss has occurred among the more valuable species.[4]

The northeast's timber industry was hit with a devastating blow in 1955 when China had reached the height of its "learning from the Soviet big brothers" and better timbering equipment was brought into the region along with other Soviet practices. At the beginning it was still a bit modest, with cutting areas limited to strips 100, 200, or 250 meters in width. But during the infamous Great Leap Forward (1958–60), the limits were pulled off cutting in order to get the most out of the machinery at hand.

The impact of those three years of the Great Leap Forward on forests in the northeast was devastating. The mentality of that era was summed up in the common phrase "eating the food left by our ancestors and creating disasters for our offspring." In 1966 Premier Zhou Enlai was heard to lament the awful effects of that policy in a discussion with leaders of the water resource and forestry bureaus: "I am very concerned especially because, in our effort to harness rivers, we have done incredible damage to the waterways and have also gone much too far in felling trees. Certainly, future generations will know this." When he spoke of harnessing the rivers the premier was referring to the problems resulting from the Three Gate Gorge Dam. And in talking about felling too many trees he was thinking about all the timber that had been taken out of the Greater and Lesser Hinggan Mountains.

Too much had been cut and not enough replaced. That situation caused President Liu Shaoqi and Vice Premier Deng Xiaoping to reach an agreement and call for "combining planting with cutting." What followed soon afterward was the chaos of the Great Proletarian Cultural Revolution; the line preached by Liu and Deng was, like everything else they uttered, all of a sudden open to intense criticism, and the new "planting while cutting" effort came to an abrupt halt. Forest research was put on the back burner in the northeast, and man-made forests grew more slowly.[5]

Though not all the disasters were man-made—in 1987 the Greater Hinggan Mountains were hit by forest fires that consumed as much as 700,000 hectares of timber—most were, and no one bothered to learn anything from them. This was the dawning of the age of reforms, and the timber-based economy was still immensely popular as a way of getting rich from the growing domestic and international demand for everything from furniture to every kind of paper and the handy throw-away chopsticks (China produces 60 billion pairs annually—as a trade-off for 25 million trees). Those trees that managed to survive the Greater Hinggan fires were simply chopped down, robbing the forests of any chance of rejuvenating themselves in any reasonable length of time.

And, yet again, the wildlife suffered the same indignities as did the trees, especially the magnificent Chinese Siberian tiger, the largest member of that family in the world. The deforestation proceeded without stopping, and many of these animals reportedly fled to the more amenable environment across the border in Russia. At the most recent count, taken in January 1999, a Sino-U.S.-Russian team of zoologists came up with the hardly uplifting estimate of only five to eight of these tigers in Heilongjiang and only ten in Jilin.

The scientists were aghast, for just twenty years earlier, more than eighty such tigers had been reported treading the forests of Heilongjiang, and even in the early 1990s there were sightings of thirty. If the animal is expected to survive inside China's borders, the very least that can be done is to stop the relentless pace of deforestation.

At the moment this does not seem very likely because the various lumber companies in the northeast, which are mostly state-owned, cannot make ends meet thanks to the 3 billion yuan in back pay that they owe to the much-beleaguered lumberjacks. The unfortunate truth is that many people in the region cannot make a living without chopping down trees. The unpaid wages of the lumberjacks are something that can be seen and understood, but what is more difficult for many to grasp is the environmental debt, also unpaid in a sense, that is being left for later generations. In Heilongjiang, the provincial forestry bureau has designated forty forest regions, but twenty-nine have virtually no growth left worth cutting.

The proud young heroes of Tiger Valley in Fengcheng county of Liaoning province, the subjects of that pop hit, had a horrifically dark vision of the future much earlier. They cut down all the mature forests and obliterated the habitat of the magnificent tiger, and in 1998 their model mountain town was itself completely obliterated by a terrible mudslide.

This horrible surprise came from an earth that they had laid bare in destroying an ecosystem that had taken thousands of years to establish. And everything they gained in several decades of logging, long by man's lifespan but short by earth's, was wiped out in a rather short period of time. It might be time to tell China's lumberjacks to stop singing those high-spirited songs—and wake up.

FROM GREAT NORTHERN WILDERNESS
TO GREAT NORTHERN GRANARY

Deer are so friendly that we knock them down with sticks,
Fish are so plentiful that we scoop them up with ladles,
Pheasants are so silly that they jump right into the cooking pot.
　　—Folk song popular among the youth who went to reclaim the Great
　　Northern Wilderness from 1966 to 1976

EVEN BEFORE THE 1998 FLOODWATERS HAD RECEDED, more than eighty members of the Chinese Ecology Association's Wetland Research Commission had issued an open letter in which they pointed out that the appalling floods on the Nen and Songhua rivers were the direct result of excessive reclamation and deterioration of the wetlands in the northeast.

To the popular mind of the time, the notion that the wetlands, especially the swamps, could be thought of as a valuable and irreplaceable natural resource was, well, difficult to grasp. To the average person swamps were dreadful mosquito-infested, vermin-ridden death traps laden with unspeakable horrors.

The reality was quite another matter, even if a bit more prosaic. Swamps and wetlands were a gift of nature that served as reservoir, life-science lab, buffer, and climate control facility all at the same time.[6] Another amazing characteristic they concealed was their ability to deal with urban wastewater, sewage, and toxic industrial waste. In view of this, some people have referred to them in a less-than-poetic manner as "the earth's kidneys." But others have perhaps taken that metaphor literally: they ate up every last bit of this valuable resource.

China's strongest and most resilient "kidneys" were mostly in the northeast. Just fifty years ago the plains of the Songhua, Nen, and lower reaches of the Liao rivers had a large number of wetlands that cleansed the river waters of much of the northeast while adding vast amounts of moisture to the atmosphere and providing a habitat for a wide variety of flora and fauna.[7]

These very fertile areas, however, were like so many other parts of China—they could not avoid destruction forever. The Three Rivers Plain referred to above was the result of alluvial deposits of the Heilong, Ussuri, and Songhua rivers in a very wide and flat area that was always referred to respectfully as the "Great Northern Wilderness." The 57,000-square-kilometer wetland was the largest and the most fertile piece of unfarmed land in 1949.

But just after Liberation in 1949, the country began establishing state-run farms in the region. More than 100 were built in just three years. The minister in charge of state farms, Wang Zhen, announced one of these farms, State Farm No. 850, with a great deal of hoopla and delivered a report to the CCP Central Committee and Mao Zedong that included a description of a comprehensive Railway Army Corps plan to establish a large agricultural complex on the Three Rivers Plain to reclaim 10 million *mu* (660,000 hectares) of "barren" land.

That report got the approval of Mao and other leaders who saw this "wasteland" as the answer to that perennial question of how to feed a rapidly growing population. After that, the state sent large numbers of demobilized soldiers and educated urban youth to open the virgin land.

Within a fifty-year period, more than 2 million hectares of the region's wetlands and swamps were turned into farm fields and more than 100 large state farms modeled on the Soviet agricultural system were established. They produced more than 100 million tons of grain and beans, at least 77 million tons of which were sent to other parts of China. Even now, local farms are still trying to develop the remaining 4.2 million hectares of grasslands, wetlands, riverbanks, and mountain slopes of the Great Wilderness, but with a new tactic: looking for domestic or foreign investors.

While the locals took a great deal of pride in this accomplishment, it began to worry ecologists. There was good reason: As the state farms expanded during that five-decade period, the wetlands shrank from 5 million hectares to only 2 million. In fact, it wasn't only the ecologists who saw the threat. The very people who reclaimed the low-lying lands got a taste of the ecological impact when they became its victims. Not long after they drained the water from the swamps, the water came back to haunt them in the form of floodwaters.

An equally, if not more, pitiful spectacle has been the loss of the birds, fish, and other animals whose final habitats were destroyed by the immigrating humans. Even an internationally recognized wetland such as the Zhalong, a focus of attempts to save the endangered red-crested crane, was pushed to the brink of obliteration when people began tapping into its scarce water resources.

Land reclamation and deforestation caused serious soil erosion around this area, which had contained some of the world's most fertile black earth. It took thousands of years for the fir forests to grow and decay to form this fertile loam, but half of it disappeared after four decades of farming. Forty years ago the black soil averaged 60 to 100 centimeters deep. It is now no more than 50 centimeters thick and in some places as thin as 25 centimeters. In the most seriously eroded areas the black soil has virtually disappeared and been replaced by barren earth.

But no matter how terrible the losses were, they only made the state farms more eager to open new lands, in the process arguing that they had made nearly 6 billion yuan in five decades of reclamation. Obviously land reclamation made good sense because it was good business.

Viewed from a very different perspective, these areas do not appear to be doing so well. On the Three Rivers Plain, overgrazing and an excessive amount of cultivation have had a terrible environmental impact, the cost of which has been estimated at 4 billion to 6 billion yuan.

There is another perspective. The damage done to Heilongjiang province by the dreadful floods of 1998 was estimated at a whopping 30 billion yuan—and that does not factor in individual or indirect losses. Some members of the Wetlands Research Commission are of the opinion that the primary cause of all this damage and indeed of the floods themselves was the loss of the wetlands. They explain that even if the reclaimed land was not the main reason for the floods but bore only a fifth of the responsibility, say, it would still mean that the people of the Great Northern Wilderness would have to work for another fifty years to pay off their "water debt."

In any case, as was mentioned earlier, when the 1998 flood hit, everything was done to protect the Daqing oilfield. But in an odd twist of fate the reclaimed land was itself reclaimed by nature, and many of the occupiers were forced to pick up their belongings and flee. And those floodwaters, which wreaked so much havoc and did so much damage elsewhere, brought some wetlands back to life. Fish could once again be seen swimming in the murky waters and birds singing while they balanced on the reeds waving in the breeze now that the swamps had been replenished. The fish, birds, and grass were back—but it's a sad comment on life that it took a devastating flood to do it.

Humans' Gift to the Rivers

Many northeastern rivers that feed into the Bohai Gulf are now open sewers and their water looks like nothing but soy sauce.
—Dou Yueming, State Ocean Administration's North Sea Branch (2001)

THE NORTHEAST LOST MORE THAN 6.6 MILLION HECTARES of wetlands and swamps in the past 50 years. It lost one of its kidneys, and water pollution got worse.

Historically, northeastern China meant, first and foremost, the Songliao River valley, which drained Liaoning, Jilin, and Heilongjiang provinces and the eastern part of the Inner Mongolia region. It included the Songhua as well as the Nen, Liao, and various rivers along the Liaoning coast. There were also the rivers that made international boundary lines, which included the Heilong (Amur), Ussuri, Tumen, and Yalu rivers.

In the twentieth century the northeast was known for its coal, oil, and lumber resources and was colonized by Japan partly for those reasons. After that, it naturally became the major industrial center for Communist China and is responsible today for at least a fifth of the nation's heavy industrial output.

Heavy industry meant heavy pollution. Hardly anyone visiting northeastern China could fail to be struck by the dense concentration of factory chimneys that dot the landscape. Many of the region's cities are at the top of the list of the most polluted cities in China or even in the world.[8]

Then there's the water pollution. More than 10,000 kilometers of the region's rivers suffer from serious water pollution, a fifth of that at Level V (raw sewage) or worse.[9] Eight major river systems in the region are heavily polluted. Only the Yalu, along the North Korean border, and certain southern parts of the Songhua and the Nen are relatively clean.

There are 50 million people living in the Songhua valley dumping 3.5 billion tons of wastewater into the river annually.[10] The greatest threat is from mercury. This may cause the reader to recall the Minamata disaster in Japan, which did not surface until the 1960s, when it was discovered that companies had dumped untreated mercury into rivers and the bay of that name a decade before. The effects were first seen in the many cats living in the area, which had a tendency to run around in a crazed manner before dropping dead. This was followed by the appearance of symptoms in humans who lived on the fish taken from the area's waters. They exhibited moments of confusion, loss of balance, and difficulty in speaking. These were the early symptoms of classic cases of neurological degeneration caused by mercury poisoning. It led to blindness, deafness, eventually a total loss of sanity, and ultimately death.

By the end of the 1970s the Songhua River began seeing dramatic increases in its mercury levels. There were 100 factories along the river that used mercury and dumped a total of 150 tons of the substance into the river untreated, laying waste to many of the river's fish. The pollution levels were exceeded only by those in Minamata Bay, and now, even after twenty years of treatment, there are still at least 50 tons of mercury precipitates in the Songhua, with ugly long-term consequences for the environment and humans.

The Liao River, which is the major water source for Liaoning province, China's heavy industrial center, is the recipient of 2 billion tons of wastewater annually from the 40 million people living around it.[11] Although the Liao gets less wastewater than the Songhua, it is actually much dirtier because it contains less than a fifth the amount of water that the Songhua does. During the dry season, it resembles nothing so much as a big sewer.

Industry is the major polluter in the region, but agriculture does its part, thanks to the abusive use of chemical fertilizers and pesticides. The polluted river returns the favor. Because of a lack of fresh water, the untreated wastewater of some cities is used to irrigate farm fields and not only enters the soil but remains on the crops as a poisonous residue. The Zhangtu agricultural irrigation district was a good example, with chromium levels on 400 hectares of fields well above the safe level. From 1980 to 1981 those fields produced 2.9 million kilograms of rice that could not be eaten. At present, the parts of that district with the worst pollution can no longer be used for agriculture. They've had to be used for industrial purposes.

The urban waste problem caused by humans and industry is shown in tables 1 and 2:

TABLE 1. Liao River Pollutants from Major Urban Areas
(millions of tons/year)

City	Discharge
Shenyang	457
Benxi	516
Anshan	286
Liaoyang	204
Fushun	169
Panjin	147
Yingkou	89

TABLE 2. Songhua River Pollutants from Major Urban Areas
(millions of tons/year)

City	Discharge
Changchun	190
Jilin	890
Harbin	290
Qiqihar	980
Mudanjiang	120
Jiamusi	560
Daqing	140

Even now, there is not a single city in the whole of the northeast with a sewage treatment plant. Each and every one of them sends its waste downstream to its neighbors, and each and every one has to accept the same donation from its upstream neighbors.

Harbin, the capital of Heilongjiang province, uses 1 million tons of water daily, with 800,000 tons of that coming from the Songhua. Harbin is on the river's lower reaches, where it is offered what amounts to a thick, turgid soup in the way of water. The water contains all kinds of dangerous substances. One study showed that the city's tap water contained fourteen different carcinogens and sixty-five different kinds of organic poisons.

In January of 1995, right in the dead of winter, water levels in the Songhua dropped precipitously while industrial pollution levels upstream were increasing, and so the city did not have an adequate supply of treatable water for either residential or industrial use. Local water authorities put emergency measures into effect, which included shutting down factories at great cost.

As these kinds of disasters occur with greater frequency, people begin looking for water underground. Historically, the cities along the lower reaches of the Liao were able to rely on its water alone. But when pollution levels rose and water quality dropped, many areas began a subterranean search. Not surprisingly, that led to a growing number of problems with funneling and cone effects.[12]

Deforestation, desertification, soil erosion, water and air pollution: When confronted with this environmental horror story, some hark back to the glories of the great northeast of yesteryear—the blue skies, white clouds, forests, green wilderness, and endlessly flowing sweet waters.

Who wouldn't want to return to that?

FORESTS TO INHERIT

How can we face future generations if we clear-cut all these wonderful Korean pines? We should leave some trees for them.
        —President Liu Shaoqi (1961)

THE NORTHEAST IS NOW LIKE THE REST OF CHINA, with more sand and more people. Before the eighteenth century the area had no more than a million people, but between 1791 and 1911 the population grew almost fortyfold. By 1949 it was 38 million; today it is more than 100 million.

124

One generation of immigrants followed another, but they now have less to inherit. In only 100 years more than half of the northeast's forests, 100 million mu (6.6 million hectares) of its wetlands, and another 100 million mu of its grasslands disappeared. Once again, it comes as no surprise that floods and droughts ravage the once-prosperous land.

In a mere fifty-year period, the land reclamation there was similar to that on the Yellow and Yangtze rivers in a much earlier period. It also turned the long sloping fields with the world's most fertile black earth into gullies and valleys redolent of the Loess Plateau. Fifty years ago, soil and water erosion touched 156,000 square kilometers of the Songhua and Liao river valleys. The eroded area has now grown to 282,000 square kilometers. In Heilongjiang, in the 1960s, there were already 5 million hectares of eroded land, but that figure is now more than 13 million hectares.

That continuous erosion increased the sediment in the region's rivers and caused a reduction in the water flow. Many of the tributaries of the Songhua were turned into "hanging" rivers like the Yellow, with sediment accumulation causing the river's bed to rise 2 meters. Its effective flood discharge capacity dropped from 7,300 cubic meters to only 3,500 cubic meters.[13] The river's famous "copper banks and iron bed" turned into sandy banks and muddy bed.

Trees limit erosion and flooding and have a beneficent effect on the weather. A study of the Changbai Mountains in Jilin province showed, for example, that forests could cause a 2 percent annual increase in the amount of rainfall; if other climatic factors were taken into account, this figure could be as high as 10 percent.

Deforestation and the destruction of swamps worked hand in hand in shutting down that massive natural humidifier, and they led to droughts. A serious drought now hits the Three Rivers Plain seven times in every ten-year period, and the Songnen Plain gets hit nine times a decade.

The systematic destruction of the northeast's rivers differed little from the destruction wrought on the Yellow and the Yangtze. In the last fifty years, while nature's reservoirs were destroyed, more than 3,000 reservoirs were built in the Songhua

River valley with a total capacity of 7.6 billion cubic meters. There were also 39,000 kilometers of dikes built.

But when nature's water retention and protection disappear, the manmade projects seem so vulnerable. Heilongjiang's more than 18 million hectares of forests alone have a total water capacity of more than 10 billion cubic meters. Then consider the 560 reservoirs built at high cost over the past fifty years. They have an 8-billion-cubic-meter capacity in all. Which is the more preferable?

Unfortunately, this is not good enough for some local leaders. They've thought that the devastating floods showed that larger dams and higher dikes were urgently needed.[14] In fact, the floods did help them get approval for a 5.8-billion-yuan dam at Nierji that could back up 8.4 billion cubic meters of water. The reservoir was thought to be capable of handling the increasing number of floods on the Nen and to reduce the flood threat downstream.

Another reason for the Nierji dam was to help transfer water from the Songhua to the Liao in the south. This north-south water transfer project plans to divert over 6 billion cubic meters of water annually from Nierji to rinse the Liao River valley. In fact, many drought-prone cities in the region have asked for permission to transfer water from other river basins.

Most rivers in the northeast still have a natural flow even when the Yellow and most other rivers of the North China Plain either dry up or have a reduced seasonal flow. But that is no reason for people in the northeast to follow in the footsteps of those people living a little south of them and start damming up and diverting rivers. All the people in the northeast really need to do is take a look at those pitiful rivers and they could understand the value of their natural rivers and keep them flowing.

On August 31, 1998, Premier Zhu Rongji paid a visit to Heilongjiang. While there, he insisted on meeting one special person: Ma Yongshun, a model worker in the area's timber industry who had by that time retired.

Ma had been a first-generation lumberjack in the "new China" and had devised a number of innovative ways those many long years ago. One of his methods allowed a single person to bring down six trees in one day. He personally met Mao Zedong and other leaders fourteen times and had the distinction of single-handedly chopping down some 36,500 trees before he laid his axe to rest. And yet, near the end of his illustrious career this model worker came to the sudden realization that the green mountains of his youth had all but disappeared. He could see it all with his own eyes. And he began to worry about the future generations when he realized that they would not inherit the natural wealth that he had. He set himself a new task: He was determined to plant enough trees to make up for all those he had cut down.

On May Day 1992 Ma and his family traveled to some mountainous areas to plant the final 1,500 seedlings that he felt he owed the mountains. Later Premier Zhu told Ma that "during the huge floods in the northeast, I frequently thought about you. Planting trees and regenerating forests so as to preserve the soil is one of the most important tasks confronting us and is crucial to maintaining the ecological balance. We must learn a lesson from this flood. We must never again cut down so many trees."

Since then, the central government has called for tree cutters to be replaced with tree planters, and it has allocated considerable money for trees in the northeast. But it would still be better to bet on northeasterners such as Ma Yongshun to restore the greenery. When each and every person living in the region can see the black soil as their greatest asset and can protect it with more trees, then future generations just may have a chance to see green mountains and clear rivers.

Black River – New Farm

Dang River – Reservoir

Dengxi – Loess Plateaux

Dunhuang – Crescent Spring

Dunhuang – Desert

Ruins of 1,800 year-old Wall and the Dying Shule

Yadan – Masterpiece of the Shule

Stone Sheep – Dying Trees

CHAPTER FIVE

NORTHERN CHINA

CIVILIZATION OR BARBARISM?

There are many wild cats in Beiping [a county in the center of the North China Plain]. General Pei Wen who was stationed there once hunted 31 of them in one day.
— *History of the Tang Dynasty* (A.D. 618–907)

THE NORTH CHINA PLAIN (HUABEI PINGYUAN) is 310,000 square kilometers in size and extends from the mountain ranges around Beijing in the north to the lower reaches of the Yangtze River in the south. It is bounded on the east by the Yellow Sea and on the west by steep mountain ranges that were a source of waters for three major rivers: the Yellow, the Hai, and the Huai.[1]

The civilization of this flat, fertile land began relatively early. Peking man inhabited the caves near Beijing some 500,000 to 600,000 years ago. At that time, and for a long time afterward, northern China was a land of great beauty, with forests covering what are modern-day Shanxi, Hebei, and Henan provinces and the area around Beijing. That forest coverage has been estimated at between 60 and 70 percent. Until a thousand years ago, it was a haven for a wide variety of animal life.

All of that natural richness departed with the arrival of a more developed civilization. When the elevated riverbed turned the Yellow River into a major "hanging" waterway, the high dikes along its banks came to define a dividing line between the Hai river system to the north and the Huai system to the south—a curious wonder.

The Hai system took in the waters of numerous tributaries and drained the plain that ran northward to the dikes of the Yellow River. It brought five major tributaries together, while a system of about 300 rivers, large and small, fanned out across the great plains. The Hai drains areas around the cities of Beijing and Tianjin and most of Hebei province and is one of China's most important waterways. *(Appendix 9 – Drainage Basin of the Hai river)*

But in spite of the system's size, the total amount of water flowing in it is very small. In fact, it accounts for less than 0.87 percent of the average annual flow of all of China's rivers. Certainly that is not much water, but even with those limited supplies the system is still a vital part of the lives of 120 million people.

And that burden is beginning to show. While a great deal of attention has been paid nationally and even internationally to the seasonal dry periods on the Yellow River, studies have shown that none of those 300 tributaries of the Hai system manages to maintain its flow year round these days. And even when they do flow, many are not rivers in the normal sense of the term, because massive amounts of pollution have turned them into something like a large sewer pipe. If it weren't for the water diversion project that took water from the Luan River to the city of Tianjin, there would not even be wastewater flowing into the city during the dry season.

The degradation and ultimate destruction of the Hai River system, especially of its once verdant forests, occurred over a period of centuries — since humans started inhabiting the area, in fact. The first emperor of the Qin (Qin Shi Huang Di) has gone down in the history books as China's great unifier and the first contributor to the Great Wall. But what is often overlooked in this record of nation building is the fact that his grandiose construction project required an enormous amount of wood. The first large-scale attack on the forests of the Yan and Taihang mountains began there.

During the following dynasty, the Han (206 B.C.–A.D. 221), a dramatic increase in the population of the empire led to large-scale land development across the North China Plain. That resulted in a reduction of the area's once-rich forests and grasslands. Subsequent dynasties had a practice of moving the capital to different cities, and the construction work on city walls and ornate imperial buildings for each and every one of them meant an increased demand for lumber from the Yan and Taihang mountains. Aside from the more obvious uses of timber for beams, supports, and rafters, it was needed for the equally important wood that fired the immense number of kilns that produced all the bricks needed to build the walls.

As Buddhism spread throughout China from the fourth century A.D. on, even more wood was needed for the many temples that still dot the area. In the Wutai mountains along the upper reaches of the Yongding River, there was one peak alone that had 300 temples, which were built at great expense for the surrounding forests.

By the time of the Yuan dynasty (A.D. 1271–1368) local forest reserves were already depleted, and the Yongding River, the largest tributary of the Hai system, began silting up so much that it soon became known to locals as the "Little Yellow River."

By the Ming Dynasty, whose capacity for destruction of the environment has already been noted several times, the population increase had pushed the land reclamation efforts farther up into the mountains. The Ming emperors made an attempt to strengthen and connect parts of the Great Wall, and by the time they had finished the large construction projects, virtually all the forests within several hundred kilometers of the wall had been denuded.

But "civilization" was not about to be stopped. By the time of the last dynasty, the Qing (A.D. 1644–1911), population growth was so unchecked that per capita access to arable land began to decline. Put another way, the ecological limits of the Hai River valley had been reached, as was made amply clear by the frequency of the droughts and floods that hit the valley. In 2,000 years of civilization, the forest cover on the North China Plain went from 60–70 percent to just around 5 percent by 1949.

An area that was known for its rich soil began to see the productivity of its land decline. As for the forests, the destruction had gone so far that it was practically irreversible. The Yongding, the Hai's longest tributary, has, according to some measurements, 60.8 kilograms of silt per cubic meter of water. That figure is larger even than that of the Yellow, where a cubic meter of water contains only 37.7 kilograms of silt on average.

But it could be seen coming a long time ago. There's an old painting hanging in the museum at the Marco Polo Bridge in Beijing's southwestern suburbs that is a vivid depiction of trunks of large trees being floated down the Yongding River by the dynastic leaders of the past. But that interesting scene bears no relation to the present. If a person stands on the stone bridge and glances off toward the Western Hills in the distance, there are no forests to be seen anywhere—they were all cut down. And the Yongding? Not a single drop of water in the wide riverbed, just sun-bleached sand and stones—it's dry as a bone.

MOVING THE CAPITAL FOR WATER

Cambaluc's large government, strong garrison, and big population all depend on supplies from the Yangtze River delta [1,200 kilometers to the south].
        —*History of Yuan Dynasty* (A.D. 1271–1368)

FOR THE PAST 800 YEARS THE HAI RIVER VALLEY has been at the very center of China's political and cultural development. But with an annual flow of only 2.26 billion cubic meters, the Hai was a fickle source of water and posed problems for several dynasties.

When people are faced with insufficient water resources their response almost never seems to be to look for a way to cut demand and preserve the vital resource. Instead, a routine response seems to be to devise all sorts of methods to manipulate hydrological systems to answer every need no matter how superfluous.

About 1,400 years ago China began digging immense canals as a way to transport grain from the Yangtze River delta to northern areas that were short of supplies. But by far the most ambitious and important canal was the one built during the Yuan Dynasty——the Grand Canal. It was the north-south connection between the Hai River system and the Yellow, Huai, Yangtze, and Qiantang. It was also the major grain conduit and, over the centuries, provided the relatively dry areas of the Hai River valley with sufficient food supplies from China's surplus-laden south, allowing the population of that region to grow beyond the natural restraints of poor agricultural conditions.

The city of Beijing (Chinese for "northern capital") was of critical importance for the region and was first built by the emperors of the Liao Dynasty (A.D. 937–1123). But it did not achieve true imperial status until the Mongol Yuan dynasty, when Kublai Khan built his capital, Cambaluc, near the site of the original settlement. From that time on, the Hai River valley became the center of a multinational empire that was continually in need of large amounts of grain.

The Grand Canal became Beijing's lifeline, and large quantities of grain were shipped up to the capital, which proved to be the center of a growing population whose demand for water grew and grew. Over the centuries this proved to be an increasingly intractable problem for the city.

When the city was first established, it relied almost entirely on the water resources in the immediate vicinity. But by the Yuan it was already having water shortages and had to divert almost all the water from springs in the Western Hills. This allowed the city, which was not near a major river, to serve as the capital of China's last three dynasties — the Yuan, Ming, and Qing.

These solutions, engineered by man, could not last. In the 800 years since the elaborate network of canals was built, man hacked away at the forests on the northern and western fringes of Beijing until the springs that fed the canals began to play out, causing the water in the system to fall a bit more every year. The desperate situation that this led to was itself the result of a number of poorly conceived water diversion projects that had taken place much earlier.

When the "new China" was founded in 1949, the new central government threw all its might into hydrological projects, especially into increasing the numbers of its beloved reservoirs. That work naturally began at home, in the capital. By 1950 Mao Zedong and Zhou Enlai had already approved the Guanting Reservoir project, which was completed in May 1954 with a total capacity of 2.2 billion cubic meters of water. It was China's first large reservoir.

It was soon followed by a major water diversion project on the Yongding River. Then there was the massive water conservation campaign that was a central part of the Great Leap Forward, which reached a fever pitch in 1958, when work on several new reservoirs in the area began. These included the large Miyun Reservoir. But in spite of its haste to complete all of these seemingly impossible tasks by 1960, Beijing still managed to suffer from water shortages. By 1965 the water problem had become a crisis, and Premier Zhou Enlai decided to increase the size of the water transfer from Miyun Reservoir to Beijing.

More water flowed into the city. But, as had been the case in the past, this simply fed the expansionist urge to such an extent that by the 1970s and 1980s the population increase had offset any benefits of the added amount of water. In the end, per capita usage did not register any increase. Then the city was hit by one drought after another during the '70s and '80s and there were drastic reductions in the amount of water in the important Guanting and Miyun reservoirs, the source of more than 90 percent of the city's surface water supply.[2]

The response to the crisis was typical—even more water retention projects. In the fifty years since the "New China" was established, the state has built more than eighty reservoirs and drilled 40,000 wells in suburbs of Beijing. Indeed, since 1949 the water supply capacity for Beijing has increased fiftyfold. Yet rather than overcoming perennial shortages, this growth in supply simply spurred more development and after that more demand both in the agricultural and industrial sectors and among an ever-expanding population.

In 1949 the population of Beijing was less than 2 million, with half that number actually living within the city walls. Today it has grown to 13.8 million people, with 7 million located in the urban area. In addition, Beijing, like many Chinese cities, is a magnet for the country's large "floating population" of day laborers. The number of those with a permanent residence permit is now somewhere around 3 million.

Another problem confronting Beijing is the same one that the Guanting and Miyun reservoirs had to face. Ecological degradation, combined with industrial and agricultural growth in the upper reaches of the streams that feed them, has led to a constant fall in the water going into the two life sustainers.

TABLE 1. Feed of Guanting and Miyun Reservoirs (M³/YEAR)

Reservoir	1950s	1960s	1970s	1980s
Guanting	1.83 billion	1.37 billion	810 million	630 million
Miyun	1.93 billion	N/A	800 million	400 million

In the 1990s the situation only got worse when the reservoirs had dangerously low water levels during several dry years. In an effort to find more water, people have been developing major projects to tap underground supplies, which has led to the systematic lowering of the water table beneath the capital and land subsidence in many areas.

One response to this situation has been to order major reductions in water use by farmers.

TABLE 2. Water for Agricultural Use at Guanting and Miyun Reservoirs (m³/year)

1960s	1970s	1980s
700 million	600 million	350 million

By 1995 only 56 million cubic meters of water were being given to local agriculture, and estimates are that sometime in the near future, this will drop to zero, as virtually the entire capacity of these two big reservoirs is devoted solely to meeting the needs of the city's residents.

Although the sharp drop in available supplies did reduce water use by agriculture, the annual consumption remained around 2 billion cubic meters. So when water from the reservoirs is denied to the agricultural sector, farmers will have no choice but to look for new supplies underground. Data show that by 1994 farm villages within the Beijing area had tapped underground water for 80 percent of their supplies.

The fact that there has been excessive demand for subterranean water in the Beijing area is well known. As early as the 1960s there was a cone of depression more than a thousand square kilometers in area already forming below the city and its suburbs. Then in the 1970s Beijing began to draw on water far from the city. In Shunyi county, for example, more than 100 million cubic meters of water from a new purification plant were diverted to Beijing.

After that, just about any suburb or rural area within striking distance that had access to water ended up being raided by the city. The result was that the water table dropped significantly. In Shunyi county it went from 6.41 meters below the surface in the 1970s to 11.40 in the 1990s; by 1995 it had reached a depth of 14.32 meters. In some areas near the county seat, it was down to as much as 42 meters. In Tongxian county, Daxing county, and many others, the water table in urban areas has dropped to more than 30 meters below the surface. Two-thirds of the city's 889,600 deep wells in fields and farms have become useless because the water table dropped too far down.

The massive use of subterranean water is emptying the city's aquifer. By 1998 estimates were that the supplies in the aquifer were 3 billion cubic meters less than what was needed. For immediate gain, man was actually laying the seeds of a future disaster.

In 1999 the dryness of the Beijing region resulted in dangerously excessive use of underground water supplies. In a six-to-seven-month period, there were serious drops in the water table from the previous year, ranging from 0.45 to 2.93 meters. In Shunyi county the water table plummeted more than 5 meters.

But the demand kept increasing. The rapid economic and population growth in the city mean that a billion cubic meters of water will be needed for the people and nearly 3 billion cubic meters for agriculture, with another 2–3 billion for industry such as the giant Capital Iron and Steel plant located to the west of the city. These three—population, agriculture, and industry—will need upward of 7 billion cubic meters, but 4 billion cubic meters is the maximum that can be taken from nature.

When amounts of water available to 120 capitals of the world are measured, Beijing ranks way down at 100. As early as the 1980s there were proposals being made that the capital be moved down south, where water was plentiful. But when this seemed unlikely to garner much popular support, other approaches were considered. These included building an even more massive water diversion project that would bring the rich water resources of the south—the Yangtze River, that is—to the arid north via a 1,200-kilometer diversion channel that would connect directly with Beijing.

This project was part of the larger South-North Water Diversion Project, which was the brainchild of Chairman Mao Zedong. The project was not approved at the time because of its astronomical cost and uncertain environmental impact. Finally in 2001, in order to save Beijing from a water crisis, the central government gave the go-ahead to the project. It is quite likely that Beijing will get billions of tons of Yangtze River water by 2010. But the experience with water diversion projects in the past has not been entirely good—they have often had bad environmental results and frequently are not sustainable. And when one considers the worrisome Yangtze situation, described in an earlier chapter, it might be wise to see to it that this diversion project does not result in another round of water consumption growth in Beijing.

NINE RIVERS CONVERGING IN A LAND OF WATER

We must find a radical cure for the Hai River!
    —Mao Zedong (1963)

THE HAI RIVER'S MANY WATERWAYS appear to form a wide fan with the megalopolis of Tianjin at the fan's fat, stubby handle. The city prospered because of the river link with Beijing and because it was the port for north China. For centuries Tianjin suffered from floods and seemed to be perpetually waterlogged because, as the old saying had it, "nine rivers converge in this land of water." Tianjin has flooded more than seventy times in the past 600 years.

China began building reservoirs and digging distribution canals in the 1950s along the upper reaches of the Hai to deal with this. The result was a series of 80 large or medium-sized reservoirs and 1,500 small ones, with a water capacity of more than 11 billion cubic meters. Meanwhile, distributary channels were dug on more than ten of its branches so that water could flow directly into the sea without passing through Tianjin. After this "radical cure" was in place, Chinese hydrologists and authorities pronounced this the first major river that had been brought under complete human control.

Unfortunately, one of the results of this elaborate planning was to cut Tianjin off from the rivers that used to converge on it. This created a water crisis, and the situation got worse in 1971 and 1972 when the serious water shortages that Beijing was itself having led the central government to shut off the Guanting and Miyun reservoir tap for the city of Tianjin.

By the end of the 1970s the Hai River was drying up frequently, leaving millions of people in Tianjin in misery. It also left less water to drive the city's electric power generators, so the city ended up with a series of rolling blackouts that hurt residents and industry both. Even the water supplies that were available to the city's residents had a saline content as high as 800 milligrams per liter. The folks of Tianjin could only joke about how their tap water was more suitable for making pickles than for drinking.

In the 1950s Tianjin had already started pumping up underground water almost without cease, a practice that caused the water table to drop and, after that, major ground subsidence. Since 1959 the amount of subsidence around the city has been about 2.5 meters on average. In the coastal district of Tanggu, at the mouth of the Hai River, annual subsidence has been 18.8 centimeters, and seawater has slowly been seeping inland. Recent figures show the water table in the city center to be as far as 95 meters down.

If Tianjin had continued on this path of pumping the underground water, ultimately it would not have been much different from people there drinking their own blood. There was no alternative; Tianjin had to look elsewhere.

Topping the list of possibilities was the Luan River to the north of the Hai. Its headwaters were in Inner Mongolia in the Zhenglan Banner. The Luan used to be clean, with plenty of water, running into the Bohai Gulf almost undisturbed. When its waters were diverted to the city, it became Tianjin's lifeline.

The diversion project was completed in September 1983, providing more than 1 billion cubic meters of water annually, and the 8 million people of Tianjin thought they could say goodbye forever to water shortages. But the Luan proved to be no more than a brief respite for Tianjin, not an eternal solution.

The reason was related to the river's upper reaches, where, as in most other parts of China, there had been considerable reclamation work for agriculture that seriously affected the surrounding grasslands and forests, some of which suffered from serious desertification. When droughts and sandstorms began to be a fairly common occurrence around the headwaters of the Luan, its water flow dropped, as did the water in lakes in the area. Along the Shandian River in the Dolonnur area, which the Mongolians referred to as the "seven lakes" area, those lakes ceased to exist. Soil erosion hit 15,000 square kilometers of land, releasing more than 40 million tons of earth every year, and rivers and reservoirs began to silt up badly. Panjiakou Reservoir, the most important water diversion project for Tianjin, has 50 million cubic meters of sediment.[3]

The destruction of the ecology along the upper reaches of the Luan in effect led to reduced capacity in reservoirs and the river itself at a time when demand for water in Tianjin was increasing. During all of 2000 and the first half of 2001, a number of reservoirs, including Panjiakou, almost dried up completely when the region was hit by bad droughts. At the height of the crisis, water supplies for Tianjin's 10 million people were expected to last only two months. That forced the central government to approve an emergency diversion plan to transfer scarce water from the Yellow River.

In a mere fifty-year period, Tianjin went from a waterlogged landscape to a drought-stricken one. And only twenty years after the massive Luan River diversion project was completed, the city's authorities found themselves on yet another water quest. At its grandest, this involved the idea of diverting water from the Yellow or Yangtze rivers and even from the ocean. Of course, the only rivers not on the list were the Hai and its branches. They used to meet in the city, but over a period of several decades they had become nothing more than sewer pipes.

CHINA'S AUGEAN STABLES

Along the big river with its wide waves,
Fragrant rice plants bend in the wind.
My ancestral home is along its banks,
I grew up to the boatman's tunes.
  —"My Motherland," a popular song in the 1950s.

AS CHINA'S RIVERS SLOWLY DETERIORATED over the past several decades, people played a curious question-and-answer game, asking which was the most polluted. If it had simply been a matter of which river had the greatest amount of wastewater and sewage in absolute terms, the Yangtze would have been the clear winner, hands down, with 20 billion tons of hideous substances dumped into it annually.

But there were those who declared the Hai River the winner. In absolute terms it had considerably less—just 4.7 billion tons—but its ratio of clear water to wastewater made it a heavyweight contender. Its ratio of 1:6.5 was one of the highest in China's seven major waterways.

Of the nine rivers and branches that make up the Hai River system, virtually none could be called clean. Nearly half of a 4,279-kilometer stretch of the waterway had contamination at level V (pure sewage). Most of the large wetlands in the area had disappeared, and the few remaining, including Baiyangdian Lake, were permanently polluted and not suitable for human use.[4]

Many of the rivers and streams near factories and mines had become nothing but waste receptacles,[5] and some of the rivers contained so many toxic materials that they corroded any exposed iron or steel parts at water control facilities in the river.[6]

In addition to the normal sources of pollution such as big factories, urban sewage, chemical fertilizers, and pesticides, a large number of new township and village enterprises (TVEs) sprang up during the economic reforms. These small-scale enterprises had little or no environmental monitoring or controls and were engaged in all manner of heavily polluting production at paper mills, tanneries, breweries, iron and steelworks, and electric power plants. In 1994 it was estimated that the Hai got 1.4 billion tons of wastewater from them, with paper mills, chemical plants, and breweries accounting for more than half of the pollutants.

When the better-off urbanites began demanding more and better meat products, the amount of animal husbandry along the rivers began increasing. This caused more pollution in the Hai system. Beijing was an example, with its 1,300 large pig farms with 260,000 pigs, its 1,000 chicken farms with more than 3.6 million birds, and, more recently its cattle lots with more than 200,000 head of cattle, all within the municipal boundaries. These livestock produce 12 million tons of waste every year, but only 3 percent is treated. Then it is either flushed directly into a network of pipes that lead to the area's waterways and ponds or left to accumulate before eventually making its way into rivers during the rainy season.

The cities and towns in the Hai River valley tap into the relatively cleaner groundwater for 58 percent of their supplies. But a recent test found that less than a third of the wells met drinking water standards. More than 63 percent of the ground water in the region is permanently polluted and unsuitable for drinking. The situation is even worse in the large cities.

To deal with this toxic assault, local governments along the Hai River have tried just about everything imaginable to protect reservoirs, which are the only remaining source of clean water for large cities. For the moment, the water of the most important reservoirs, such as Panjiakou and Miyun, is still within the acceptable range. But in view of the enormous quantities of wastewater and sewage being dumped into the drainage system annually, these are clearly under threat. Some major reservoirs that provide drinking water for several million people have already become contaminated. What they have is acres of dead fish, dead crayfish, and a dead river feeding into them, with water so corrosive that even the reinforced concrete gets eaten away. It is difficult to imagine human beings surviving under such conditions.

In the 1970s there were three major outbreaks of pollution in the Hai system that caused the nation to sit up and take note. These were at Baiyangdian Lake, mentioned above, at Guanting Reservoir, and in the Ji Canal near Tianjin.

Guanting was the first large water retention facility built after 1949 and for years was a major source of water for Beijing, with a good quality rating. But in 1971 things took a turn for the worse and the reservoir's water began deteriorating. It took on a distinctly yellowish hue, decorated with large patches of white foam, and had a bitter medicinal taste. Dead fish could be seen floating on the surface in greater numbers, and people who drank the water or ate the fish began suffering from bouts of vomiting or nausea or exhibiting stranger symptoms.

The Ji Canal, near Tianjin, had been known for its rice paddies and fish products. It took a turn for the worse in the 1970s, when large amounts of wastewater were being discharged into it. Almost overnight, many fish and other aquatic life disappeared. Those fish species that did manage to survive had a distinct odor of DDT. The water contained, on average, 11.88 milligrams of mercury per liter.

In the 1980s the Hai River deteriorated so badly that the pollution caused 425 million yuan worth of damage. The situation got much worse in the 1990s, when pollution was responsible for 4 billion yuan worth of damage every year.

The wastewater and sewage were extremely harmful. A provincial study of two villages in Hebei found that in one,

Nanjiao, where pollution levels from the nearby Jiao River were quite high, nineteen people died of cancer in a five-year period, accounting for almost 30 percent of the deaths from unnatural causes. In the second, Chenglang, where pollution levels were low, eleven people died of strange diseases, accounting for 21 percent of the deaths from unnatural causes. The normal life span in Nanjiao was 64.7 years. In Chenglang it was 74.7. The cause seemed to be fairly obvious.

That did not stop people from dumping wastewater and sewage into rivers. In more recent years, in many places along the Hai River people have no longer been content simply to dump wastewater into the river's lower reaches but have turned instead to pipeline projects that allow them to divert their waste into neighboring provinces and counties. This has led to conflicts.

The Wei Canal, which is connected to a number of rivers, has become a conduit for wastewater from four cities in Henan. So neighboring Hebei province stopped allowing the canal's waters to flow into its territory. Authorities in the city of Tianjin and in Shandong province took the same steps. The result has been that Henan's wastewater often sits stagnating at the Sinu sluice gate at Dezhou, with no place to go.

Along the lower reaches, people mostly have no choice but to take the pollution from upstream. Sometimes they get some compensation from the polluters, but they are still left to live with their own altered crops, fish, and animals, and their failing health.

The Tuhai River, which is a major source of water for northern Shandong, has seen pollution levels increase since the 1980s, leading to a serious decline in water quality and a negative effect on the lives of people living along its banks. The pollution primarily comes from pulp mills, tanneries, and breweries along the upper reaches. The area above Zhanhua is one example, with twenty-three large mills pouring out 21 million tons of waste annually, which flows past the city. The river water is toxic, and people who drink it suffer from extremely high rates of intestinal diseases and cancers.[7]

The carcinogens in the Hai River valley are some of the most alarming and are, according to many people, the main cause of the shockingly high cancer rates. In some areas the general decline in public health is blamed on poor water quality.

Conditions are so serious that these areas can no longer meet their quota of young men for military service.

The simple fact is that these people need water and are left with no choice but to use whatever is available, especially for irrigation. In the Beijing area, 233,000 hectares of land are irrigated with 180 million cubic meters of water with a high waste content.[8] A study done in Tianjin found that as a result of the long use of wastewater for irrigation, carcinogens had entered the food chain. There was also a big jump in the number of parasitic diseases and intestinal disorders that led ultimately to high rates of cancerous tumors.

Another study of the effects of using wastewater for irrigation found that out of 30,000 children below the age of five, 5.4 percent suffered from acute diarrhea. That was 1.4 times the figure for areas where wastewater was not used for irrigation. Acute diarrhea accounted for 2 percent of the deaths in the former area, while in the latter it accounted for almost zero deaths. Overall, data from the State Environmental Protection Bureau have shown that the Hai River valley has 667,000 hectares (100 million mu) of land being irrigated by wastewater, or 10 percent of the agricultural land.

We never hesitate to sacrifice habitat, animals, and everything in nature just to protect the holy lives of human being. When a river starts to damage the health of so many people, it is obviously on the verge of collapse or has collapsed.

SAVE WATER, SAVE HAI RIVER

Southern China has too much water and the north has too little. We should try to borrow some water from the south to help the north.
　　—Mao Zedong, 1952

THE SOUTH-NORTH WATER DIVERSION PROJECT, first proposed by Mao in 1952, had been shelved for nearly fifty years until it was dusted off in 2001. The central government has announced that it is ready to begin building two of three routes to pump water from southern rivers to the north.

Of the two approved routes, the east one plans to pump 13 billion to 17 billion cubic meters of water a year from the end of the Yangtze River through the Great Canal, dug in the Yuan Dynasty. The water will rinse northern provinces such as

Shandong and Hebei. Its primary target is to relieve Tianjin of its chronic water shortage. The total cost is estimated to be between 53 to 60 billion yuan.

The middle route of the project aims mainly at Beijing, but it will also help the Hai River basin provinces of Henan and Hebei. According to the plan, a 1,241-kilometer canal will be dug to divert 11 billion to 14 billion cubic meters of water annually from the Han River, the longest tributary of the Yangtze River. It will cost 78 to 88 billion yuan.[9]

Millions of residents on the North China Plain were excited by the news, but some environmentalists pointed out right away that the project will cause serious environmental problems. The east routes, set to draft water from the end of the Yangtze, they said, would worsen the salt water intrusion in the dry season and therefore hurt the water quality in the Yangtze River delta region. It could also cause sediment buildup in the mouth of the Yangtze, they warned.

Chinese environmentalists believe that the middle route will have much a bigger impact on the environment because the source of this diversion project has highly limited resources that can barely support the fragile ecological balance. These experts warned that taking too much water from Han River will worsen the pollution in its lower reaches and threaten the quality of drinking water for millions of residents in Wuhan, Hubei province. But to most water officials and experts who know how bad the water situation is in the Hai River valley, the potential troubles in the south are too trivial to be considered.

According to a survey conducted in 1993, of the forty-two cities located in the Hai River valley thirty-eight suffer from serious water shortages. We have discussed the water crisis in Beijing and Tianjin in the previous sections of this chapter, but they are by no means the worst cases. As large metropolises, at least they have the privilege of grabbing water resources from smaller cities and the countryside when they run out of water.

When most of the rivers are dammed and kept for the use of large cities, smaller cities and the peasants are forced to use underground water. This has led to a rapid decline of the water table in the plain regions of Hebei that surround Beijing and Tianjin.

TABLE 3. Average Water Table Depth in Hebei Province (meters)

1980	1985	1995	2000
4	7	10	13

The most severe reductions in the level of underground water tables in Hebei have taken place in the cities of Xingtai and Handan, both in the southern part of the province, and in the central provincial city of Shijiazhuang.

TABLE 4. Depth of Water Table in Hebei Cities (meters)

City	1978	1995	2000
Xingtai	7.46	21.33	28
Handan	4.79	17.32	24
Shijiazhuang	5.25	15.67	25

Nor is there much room for further utilization of water supplies in Hebei, as estimates are that of the 60 billion cubic meters of nonreplenishable deep groundwater reserve in Hebei, half have been emptied. The huge cones of depression in Hebei province, Beijing, and Tianjin have joined with those in northern Henan and western Shandong to create a "superfunnel" covering an area of 40,000 square kilometers, the largest of its kind in the world, said the latest groundwater report issued by the Ministry of Land and Natural Resources in 2001.

Water experts estimate that the groundwater reserve in some cities, including Xingtai, Handan, and Shijiazhuang, can last no more than fifteen years. Considering such a dire situation, no one should be surprised to see the Chinese government revive the South-to-North Water Diversion scheme despite its potential environmental hazards.

In a way the water diversion project is an emergency method to prevent the collapse of the Hai River drainage system. Just like the cases of many other south-to-north canal projects implemented in the past 1,400 years, we have again cornered ourselves with population growth and economic development that cannot be supported by native rivers.

But the advocates of the engineering solution do not acknowledge that the project is only an emergency solution. To them this project provides essential water resources needed to restore the damaged ecology of the Hai River drainage and therefore is a long-term solution.

However, when we examine the issue retrospectively, we find that all the engineering projects only increase the ecological damage in the region. Again, the reason is simple. In designing the ancient transportation canals, our ancestors only calculated how many more people the cargo grain could feed. In designing any one of the mammoth reservoirs, we just calculate how much electricity it can generate and the number of factories it can support along with the amount of land it is capable of irrigating. All of this leads to a spiral of growth and expansion that effectively robs the ecosystem of its ability to sustain itself and worsens the ecological degradation.[10]

When billions of tons of Yangtze water finally reach our doorstep, we can well imagine another round of economic expansion and population growth that further squeezes the space of mountains, rivers, and all other creatures in the Hai River region. Besides, drafting so much water from the south will inevitably negatively impact the fragile environment in the Yangtze River region.

Now we seem to fall into a dilemma: We must divert water to ease the imminent water crisis, but the diversion will hurt us in the long run. The only way to extricate ourselves from the situation is to reform the financing system of large water projects. For decades most of our large projects have received investment from the central government. Therefore local officials and people tend to build them as large as possible. Many such larger-than-necessary projects cannot even cover their operating cost, let alone recover the investment.

To prevent the water diversion project from becoming another huge financial burden, we should make the end users cover a substantial part of the cost of the project, if not all of it. The true cost of water diverted to Beijing is expected to be higher than 6 yuan per cubic meter. We need to raise the water price substantially right away and finance the project with the surplus income.

The biggest advantage of the new financing mechanism is to promote saving water in this region. As we discussed in the Yellow River chapter, higher water prices will force farmers to give up water-guzzling crops and encourage more efficient irrigation. It will also force factories to recycle water. In the special case of the North China Plain, it will help check the overexpansion of some high-water-consuming industries. Currently the region produces 20 percent of China's steel, 10 percent of its power, and 14 percent of its paper, using huge amounts of water and causing severe pollution.

Not only will higher water prices make sewage treatment feasible, they also are an efficient way to prevent millions of people from flocking to drought-hit cities such as Beijing.

Finally, this is perhaps the only way to control the scale of the diversion scheme and therefore minimize the impact on the Yangtze River.

Again, we have seen some encouraging signs here. When Premier Zhu Rongji gave his go-ahead to the project in 2001, he actually placed three conditions on it, saying that the beneficiaries needed to try all possible means to save and recycle water and to improve the environment beforehand. The water price in Beijing, according to municipal government sources, will be raised gradually from the current 2 yuan per cubic meter to 6 yuan by 2005, apparently in preparation for the project.

China's water minister, Wang Shucheng, has gone even further, suggesting that the project be corporatized and that greater water rights be given to those who invest more. To prevent the project from becoming another immense boondoggle, the minister wants each city to sign a minimum water use contract for the Yangtze.

Meanwhile, many urban residents are unhappy with the price hike, and most local governments of recipient cities still refuse to foot the bill, insisting that this should be a socialist welfare project. Some agricultural experts also warn of a great drop in China's grain output if the price of water goes too high, saying there is no way for farmers to accept anything higher than 0.3 yuan a ton.

No matter what methods we adopt, Beijing residents have almost been assured of water relief by 2008, when the city hosts the Olympic Games. And those in Tianjin will get the water from the south even earlier. Some people believe that we do not need to be bothered to salvage the Hai River; we can carry on without it.

But history tells us that water diversion is no permanent solution. Let's take the water diversion as an emergency method and build and run it in an environmentally sound way so as to save water, save the Hai River, and in doing so save ourselves!

"THE HUAI RIVER MUST BE HARNESSED"

Thousands of people have no way to escape. Some have climbed trees but then fell into the water and drowned. Others were bitten by poisonous snakes that have also sought safety in trees. Some have climbed onto boats that were quickly capsized by the rush of the flood waters and the huge waves.
　　　　—Zeng Xisheng, then governor of Anhui province, reporting to Mao
　　　　on the floods in 1950

THE NORTH CHINA PLAIN IS SPLIT INTO TWO HUGE AREAS by the Yellow River dikes: To the north are the plains of the Hai River system and to the south is the Huai River valley, two-thirds of which also consists of flat plains. Here there are more than 120 million hectares of arable land and a population of over 150 million people, one-eighth of the entire nation. Flowing through the middle of the country, the Huai River is a natural demarcation line between China's north and south.

The Huai River is approximately 680 miles long, with numerous tributaries feeding into the mainstream from all sides. Until the twelfth century A.D., the Huai flowed unimpeded and rather calmly into the Yellow Sea. But the ecodestruction in the middle reach of the Yellow River finally caused it to change course and take over the Huai waterway for 650 years. The Yellow River brought 1 trillion tons of sediment into the Huai in those years, elevating its riverbed and eventually blocking it from directly flowing into the sea. Instead its waters now flow south into the Yangtze. It dramatically reduced the effectiveness of both the mainstream and tributaries in flushing out dangerous floodwaters converging from tributaries from both sides, often leading to devastating disasters.

In response to its often unruly behavior, the Huai River has throughout Chinese history been the target of one Chinese leader after another who has put "harnessing the Huai" (*zhi Huai*) on their agenda. But still the history books are full of terrible notes on how the Huai River floods drowned thousands of people and left millions homeless. In 1950 the leaders of New China decided that despite the many failures of their predecessors, they would now succeed in "harnessing the Huai."

July 18, 1950: In the midst of the disastrous flood spreading throughout the Huai River valley, Mao Zedong, chairman of the newly installed CCP, gave his second in command, Zhou Enlai, instructions saying that "in addition to curtailing the current flood, the premier should seek ways to solve this problem once and for all. We must begin work now so that by the fall a large-scale project for diverting the Huai River can be in place and completed in a year so that the same kind of disaster does not recur next year."

Mao's instructions put Zhou Enlai in something of a quandary. The year 1950 was the beginning of "new China," and the state treasury was virtually empty, with hardly enough funds to provide for current disaster relief. Where would he find the money for such a large-scale project? Besides, the project had to get the approval of the leaders of all the provinces that would be affected, that is, Jiangsu, Shandong, Henan, and Anhui. Already Jiangsu's leaders had made it clear that starting up such a project within a year's time was virtually impossible.

Of course, Mao Zedong always had a way of bringing about a unity of thought among senior-level cadres. According to a recently published memoir by Ms. Qian Zhengying, the former minister of the Ministry of Water Resources, "Chairman Mao talked about Chinese history and underscored that the Huai River valley area was the place where peasants organized uprisings against existing regimes and from where many new emperors had arisen. For instance, Liu Bang was from Pei county and Xiang Yu from Suqian, and Zhu Yuanzhang (also known as Ming Taizu, the founder of the Ming dynasty) from Fengyang—all in the Huai River valley. Given that the region is very poor and is often visited by the natural disasters that spell trouble for a sitting government, peasants in this area are known to rebel, especially when driven by disaster and hunger."

And so the necessary funds were allocated by the Central Committee for major investments in a series of water control projects that were constructed along the Huai and its tributaries.

In addition, the riverbed was dredged, dikes were reinforced and raised, and numerous reservoirs and water diversion and retention projects were constructed, including a canal that channels water directly to the sea.

The high efficiency in building water projects in the Huai River valley convinced Mao that the economy could make a Great Leap Forward by organizing a mass movement. When this new water control campaign swept through the Huai River valley like a hurricane, Chayashan, located in Xinyang prefecture, became the first so-called People's Commune personally anointed by Mao. Such preposterous slogans as "Let the high mountains bow their heads" and "Let the river waters yield" were introduced into the Chinese vocabulary by people working at the various construction sites involved with "harnessing the Huai River." Between 1958 and 1960 eighteen additional reservoirs were constructed on the Huai with a total water retention capacity of 4.8 billion cubic meters.

But the result of all this sweat and blood and expenditure of vast sums was not prosperity but rather devastating famine. That the Huai River valley was not spared devastation is hardly surprising since many of its provincial leaders had been early advocates of the Great Leap Forward. In the wake of the disaster, dead bodies could once again be seen everywhere in both Anhui and Henan provinces. And in Xinyang, Henan, where only two years previously residents had bathed in the glory brought about by the establishment of the first People's Commune, the number of people who died of hunger was at least 200,000.

But the efforts to harness the Huai did not stop. Thirteen more large dams were erected on the upper reaches of the Huai, and the total water retention capacity reached 9.4 billion cubic meters by the mid-1970s. With a comprehensive water control system, the Huai was believed to be harnessed once and for all. Unfortunately, a greater disaster lay ahead for the Huai River valley.

August 1975: A typhoon crossed over into Henan with devastating effects. In three successive deluges that inundated the headwater valleys of the Huai River, Zhumadian prefecture in Henan province was especially hard hit. By dawn on August 8, flood waters that had been building for days on the Ru River (a tributary of the Huai) breached the large-scale Banqiao reservoir, causing catastrophic failure. It started as a water wall several tens of meters high and then drowned the more open and densely populated regions under water 4.5 meters deep.

At the same time, the Shimantan reservoir on the nearby Hong River also collapsed, producing floodwaters from the two reservoirs of over 12 billion cubic meters that swept away at least a dozen smaller dams and overwhelmed several catchment areas and dikes, rushing all the way toward the Fuyang region. After more than twenty days of massive rescue efforts involving contingents of the People's Liberation Army, the situation was finally brought under control.

In Henan province, the flood inflicted damage on twenty-three counties and municipalities with a total population of 8.2 million. In all, the final official death toll for Henan was 25,000. In nearby Anhui province, losses were also significant, as 4.5 million people were affected and 990,000 homes and buildings were damaged.

Beyond the enormous costs of the 1975 dam collapse, at present there are still some 5,300 dams and reservoirs on the Huai River that must still shoulder the enormous task of flood control and provide for irrigation. Yet many of these facilities are in poor condition, even decrepit and in need of immediate reinforcement:

February 11, 1998: The managers of three large-scale reservoirs in Anhui province convened a meeting to discuss measures for "saving aging reservoirs in China." They pointed out that the three largest reservoirs in the Huai region, Foziling, Meishan, and Xianghongdian, which were built hastily during the Great Leap, were all quite "sick" and could collapse in big floods.[11]

Most worrisome was that together these three dams have a water retention capacity ten times greater than that of the Banqiao and Shimantan dams, which collapsed in 1975. And given their strategic location, the lives of millions of people spread out in ten counties in the two provinces depend on these structures holding. If anything were to happen à la 1975, the catastrophe would be unimaginable.

In reality, the condition of these dams is emblematic of the current situation all along the Huai. Statistics indicate that of the thirty-five large-scale water retention facilities on the river, only twelve are capable of safely playing their critical role in flood

151

prevention, while a mere four could hold up during a so-called thousand-year flood. The rest are below acceptable standards and could also be characterized as "sick." But given their role in generating electricity, they continue to function.

Despite this concern over floods, the basic fact is that the volume of water in the Huai continues to drop. The entire Huai system has seen an increase in droughts that have grown from the levels of the 1950s, when 1 million hectares of land were affected by lack of water, to those of the 1980s, when 1.84 million hectares suffered from inadequate moisture. The fact is, however, that this drop in water levels has not eliminated the threat of floods and subsequent waterlogging. As the locals describe it, "small floods bring huge losses." By the 1990s the Huai had seen it all: floods, waterlogging, drought, and more pollution.

Just how much can this old river take?

June 1991: Once again floods hit the Huai River valley, bringing destruction and dislocation to 54 million people. One year later, people on both banks of the Huai who survived the summer of 1991 were hit with a major drought that became the beginning of a new chronic nightmare.

"TOUGH GUYS SEVER THEIR LIMBS"

The headwaters of the Ying River are crystal-clear.
　　　　—Li Bai, Tang Dynasty poet (A.D. 701–762)

IN EARLY 1992 ONE OF THE MANY TRIBUTARIES OF THE HUAI RIVER, the Ying, was struck by a major pollution spill in the form of 150 million tons of wastewater and sewage. The impact on the Huai was immediate, as massive amounts of waste floated into the main stream between the cities of Huainan and Bengbu, where the river's waters are drawn upon for both human consumption and industrial use.

Statistics show that the daily release of industrial wastewater and human sewage into the waterway of the Ying is 1.3 million tons and 450,000 tons, respectively. None of it is subject to any treatment before being discharged into the river, including the untreated output from paper pulp mills, chemical fertilizer plants, and other equally destructive toxins.[12]

Two years later, in the summer of 1994, another case of serious water pollution occurred, once again involving the Ying River. At the very moment when the mass of waste and sludge was pouring into the river, the Ying River valley was suffering a severe drought in which virtually no rain fell for days, causing the level of fresh water in the river to drop precipitously. With 100 large and small reservoirs and 300 sluice gates on the Ying River, local people cut the river off completely to keep the maximum amount for their own irrigation use most of the time. The result was that large areas of the Ying became virtually stagnant and turned into a smelly cesspool.

By mid-July Henan finally saw significant rainfall, and the reservoirs and sluice gates upstream and downstream began discharging water. The mass of polluted water flew down the Huai River with unbearable smells and odors. Trees in Shenqiu Park, near the last major sluice gate, withered almost overnight, while monkeys kept in the park were virtually blinded by the filth-laden air. In all, a pollution belt of more than 100 kilometers in length and consisting of 200 million cubic meters of wastewater was floating quietly downstream toward the already dried-up lower reaches.

Dead fish were seen floating on the river surface in Fengtai county, Anhui province. Huainan municipality tried to cope with the impending crisis by adding new purifiers to its city water facility, but to no avail; city residents saw a yellowish brown liquid when they turned on their tap water. Those careless enough to drink this ooze were immediately taken sick, and many had to be hospitalized.

Shortly thereafter the pollution belt burst through the Bengbu sluice gate, creating a turgid flow that could be smelled for miles around despite efforts by a local water purification plant to bring it under control by dumping activated carbon into the stream. The 500,000 residents of Bengbu suddenly confronted an immense crisis in their daily water supplies, leading to a pathetic scene where in a town sitting next to a major river, people had to buy high-priced bottled water from street vendors. Once the immediate crisis had passed, the local purification plant tested for water quality and to its astonishment found 95 major pollutants out of a total of 129, including 67 carcinogens.

Further downstream in Xuyi, Jiangsu province, the town was left with no choice but to seek supplies from underground sources. Yet even after drilling 150 meters into the ground, still no water supply could be found. The reaction among the population was to engage in their own version of a water war by fighting over access to the town's limited supply.

As for the economic impact of this increase in river pollution, the loss in the fishery industry in Hongze county alone was 200 million yuan. In addition, factory shutdowns because of excessive pollution produced losses of 220 million yuan in cities such as Huainan, Bangbu, and Xuyi, where rates of serious illness also rose dramatically.

Unfortunately, the Ying River is not alone as a major contributor of filth and waste into the Huai River system. Sources of pollution are found from the very beginning of the river to its end and along all the tributaries that feed into it. In Henan province alone there are 1,300 small but highly polluting paper mills. Henan is also known as the kingdom of small-scale tanneries, another major contributor to water pollution.[13]

Situated along the Huai River are 128 counties and 43 municipalities, yet there is not a single wastewater treatment plant in any of them; virtually all raw wastewater and sewage is dumped into the river directly. Data indicate that human and industrial wastes discharged into the Huai River from the four provinces of Henan, Anhui, Jiangsu, and Shandong total 2.3 billion tons annually. The effects? In Anhui province, along the 1,506-kilometer stretch of the Huai River traversing the province, 89 percent of the waterway has pollution registering at levels IV and V.

The pollution disaster finally stirred up the central government, and Song Jian, a member of the State Council in charge of environmental protection, took an inspection tour of the Huai in May 1994. He saw not only dead fish floating on rivers but also badly tainted wells and local villages and towns struck by extremely high rates of cancer. When blocked by thousands of protesters in Shenqiu, the shocked minister declared, "If we are not determined to clean up pollution, we cannot face the 100 million people currently living in the Huai valley nor their offspring; indeed, we cannot face history!"

On August 8, 1995, the State Council promulgated China's first river valley environmental protection law—Provisional Regulations Governing Water Pollution Prevention and Harnessing of the Huai River Valley—which required that by the end of 1997 all industrial pollution along the waterway must meet minimal standards issued by the government and that by the year 2000 virtually all sources of such pollution must be cleaned up, creating a "clear" Huai River.

Unfortunately, it goes without saying that it is easier said than done to solve the problems of water pollution by closing factories that fail to reach the standard. The fact is that over the course of the economic reforms put into place since 1978, innumerable township-owned and collectively owned factories have been set up in this region with little or no pollution control equipment and continue to exist by the sweat and hope of the peasantry who look to these facilities as their future. So any massive shutdown is an extremely difficult task. Caught between a rock and hard place, Song Jian was, however, very resolute; indeed, he took the position that to save the Huai "we must be willing to sever our limbs when bitten by a poisonous snake!"

As much of the nation watched, 1,562 factories along the river were obligated to comply with standards by midnight on December 31, 1997, or face being shut down. Hundreds of factories were closed in the high-profile campaign, and the rest—1,139— were said to meet the standards. This became the first of a series of "zero o'clock actions" taken in China.

Three years after the cleanup campaign, the Huai River is almost as polluted, with recurring pollution accidents. Near the Shenqiu sluice gate, where the monkeys were blinded by pollution discharge, primary-school students have to put on gauze masks in classrooms now and then when the sluice gate is lifted.

The truth is that people in this underdeveloped region found that they could not treat all the wastewater they generate because the cost is higher than the value they generate in production. Estimates show that the total cost of cleaning up the Huai River would come to over 1,000 billion yuan, yet the value of total annual output in the entire Huai River valley is currently only 170 billion yuan.

Obviously, without changing the mode of economic development in the region by becoming more accommodating to the natural ecology, the Huai River's water shall never be clear.

LIMITS TO GROWTH

Human beings are at the center of concerns for sustainable development. They are entitled to a healthy and productive life in harmony with nature.
—*Rio Declaration on Environment and Development,* 1992

AFTER 1949, MAINLY TO PUT THE FLOODS UNDER CONTROL, 5,100 large- and small-scale reservoirs were constructed along the upper reaches of the Huai waterway, with a water retention capacity of 25 billion cubic meters. In the middle reaches, more than 10 major flood control facilities and retention reservoirs were built, with a total capacity of 34 billion cubic meters. Previous sections have shown that these measures failed to control the flood disasters of the Huai River valley. Meanwhile, pollution has become a rising problem over the past twenty years, and droughts have also been a lingering problem of the region ever since the Yellow River ruined its water system.[14] But these gigantic reservoirs, plus more than 500 large and several thousand small-scale sluice gates and irrigation and pumping stations constructed on the rivers, must have freed the region from the threat of drought. Or have they?

Data available actually indicate a worsening water shortage in the region. In years of average water flow, the whole valley is short some 3.8 million cubic meters of water, while in dry years the deficit is as high as 8.4 million cubic meters. During the major droughts that hit in 1986, 1988, 1991, 1992, and 1994, the severe reductions in water caused enormous economic losses for the entire Huai River valley. From 1999 to 2001, which were three consecutive years of droughts, the Huai River dried up even in the rainy season, a phenomenon previously rarely seen.

And yet the more hydro projects that are built, the more conspicuous becomes the water deficit problem, which leads to the question of whether the two are in fact connected. Many people find it hard to accept such a notion, and they cite the following facts: The comprehensive system of flood control, irrigation, and water diversion built on the Huai River system has provided 1.3 trillion cubic meters of water to factories and cities in the valley, and helped put more than 13 million hectares of land under irrigation beginning in the 1950s.

Along with the economic progress have come more and more people, and by 1989 the population of the entire valley was 150 million, or 537 people per square kilometer, a figure 1.5 times the national average. To be sure, without the huge Huai River water control system, the valley would not have sustained so many people. Yet is this style of development itself sustainable?

These hydro projects have provided favorable conditions for economic development, yet the fact is that more development makes those projects less than adequate. Thus new projects are launched to compensate for the inadequacy. In what has become a kind of vicious cycle, the water shortage in the Huai valley has become so severe that many experts look to the diversion of Yangtze River water into the Huai via the South-to-North Water Diversion Project as the only viable long-term solution.

Hence the development of the region has followed the mode of projects producing development that, in turn, calls for more projects. Desire for development is infinite, but water resources are finite, and so sooner or later the lack of sufficient water will set inevitable limits on growth.

To make it worse, reckless development with little care for nature makes the limits come to us much earlier. To seek industrial gains we have turned the river into a poisonous stream. To grow enough grain for all the surplus population we have intruded upon the territories of the rivers and lakes and the once-lush hills.

In the Huai River valley, the 1.46 million hectares of hillside land that have been developed for farming have led to serious soil erosion. At present, in the entire valley, soil erosion has affected 74,000 square kilometers, 27 percent of the total land area in the valley, with annual erosion amounting to 260 million tons. More than 150,000 hectares of land has become so eroded that it has been reclassified as "barren," and more than 10 percent of reservoir capacities have been silted up.

When we finally are forced to confront the limits, the economy will be badly hurt and the population will be reduced. It will bring humanitarian disasters; the droughts that killed 2 million people in East Africa were one example. Just to avoid such disasters, we have to adopt the painful practice of family planning. But when checking the population, we may also need to find more efficient ways of living and/or economic development.

Among many Chinese there is the common belief that if the country's huge population of 1.2 billion could be halved, then the average person could enjoy a standard of living on a par with Americans. But the fact is that this notion is probably false, for it is true that with its emphasis on consumption, the United States, with less than 300 million people, has also wreaked great havoc not just within its own borders but on the world. Smaller numbers, in other words, do not necessarily mean a more viable environment. More realistically, China cannot just do away with 600 million people, and so some other method must be found for restraining ourselves and learning how to coexist with others in a relatively small space.

More and more people in the world are beginning to accept the concept that economic growth itself is not the ultimate goal of human beings. The growth should not hurt our right to live in harmony with nature and enjoy life itself.

It is to be hoped that one day the denizens of the Huai River valley will learn that they cannot abuse the river for the sake of unlimited economic development and that they must adopt a developmental style that is sustainable. That day, however, will come only when people realize that the most valuable thing in the world will be a handful of clear water from the Huai River.

CHAPTER SIX

SOUTHEAST CHINA

"WATER FLOWING AMIDST HOUSES UNDER SMALL BRIDGES"

Paradise is above, but down below we have Suzhou and Hangzhou.
—Chinese folk saying

MANY CHINESE WOULD PROBABLY SAY that the plain of the Yangtze River and its delta are the essence of China. Although these lie at 30 degrees north latitude—a level at which many other parts of the earth are desert—they are crisscrossed with a vast network of canals, rivers, and lakes. One of the most famous and important of these is Lake Tai (Taihu) in the center of the plain. But the lake itself is surrounded on almost every side by a vast series of other lakes.

The Chinese began growing rice in the fertile land around those lakes 7,000 years ago. Farther north from this part of China, the environmental damage started getting bad as early as 1,000 years ago, and the damp, warm weather of these plains meant that they became China's major food supplier. But then people began draining the lakes to make more cropland. After that, the floods got worse.[1] (*Appendix 10 — Lake Reclamation in the Ming Dynasty*)

Still, until about thirty years ago, the plains were a scenic area. There were towns where the water flowed gently among the houses, along streets, and under graceful arched bridges, giving the Chinese of those early years a poetic sense of life with soft breezes, flowers, hills, water, clumps of bamboo, fish, birds, grass, and even the song of insects. Of the many "water towns," there were two that stood out—Suzhou and Hangzhou. These were the real prizes because they exemplified the degree of harmony with nature that human beings could attain.[2]

Hangzhou was famous for West Lake, while Suzhou, just to the north, was known best for the Venetian magnificence of its canals meandering everywhere. But it was distinct from Venice or Amsterdam because its vast network of canals contained only fresh water. And it stood out thanks to the sense of timelessness imparted by the age-old stone steps along the canals leading down to the water where women would wash their vegetables and rice while their children romped about nearby in the company of large numbers of fish visible everywhere in the clear water.

These delicate scenes are now, unfortunately, consigned only to faded photographs or the memories of people old enough to remember those days. But, the degradation of the water took place right before their very eyes, often just outside their front door. In reaction, people covered their sense of unease by composing ditties like the following: "In the '50s, the water washed vegetables and rice; in the '60s, the water became worse; in the '70s, the fish and crayfish disappeared; in the '80s, the water couldn't be used even to clean a toilet."

Behind the changes lay the "revolution" in agriculture that changed forever the traditional way of farming. For centuries, to support the double cropping the region was known for, people collected every bit of organic waste they could find and put it in a compost heap to let it ferment for fertilizer. Nothing was wasted, and even human waste, or "night soil," went into "honey buckets" all across town for transport to the fields.

Every winter and spring the farmers would dredge the nearby rivers and canals and add the sediment to the fertilizer. They cut the dense grass at the water's edge to add to the pig fodder that consisted of rice and wheat stalks as well as other leftover vegetables, which, after being digested by the pigs in their sties, produced manure.

The entire process of recycling was labor-intensive but efficient in its own way—and organic. That was how the rivers and lakes were kept relatively clean in the delta, even after at least a millennium of intensive farming.

This all changed about twenty years ago with the introduction of chemical fertilizers. Human feces now were neglected and ultimately ended up being dumped untreated into rivers. And because there was a plentiful supply of chemical fertilizers, farmers no longer felt the need for the labor-intensive dredging of sediment from the rivers.

They also stopped scooping up the barnyard manure and just dumped it all directly into the waterways. At the same time, they stopped using the rushes and reeds and turned to processed feed for their animals, so the waterways eventually ended up being clogged by the unchecked plant growth.

During the same time period, to add insult to injury, there were the other effects of rapid industrial development in the post-1978 era of economic reforms. Certainly, however, it was agriculture rather than industry that was really to blame for the fundamental change.

Beautiful Suzhou could not escape the effects of this. The beginning was simple enough. The farmers stopped carting off the waste from the public toilets and dredging the silt out of the city's canals. Then people started treating the dirtier waterways as a handy trash bin where they could dump just about anything. Local statistics show as much as 15,000 tons of trash going into the waterways every year.

In many of these water towns, because the government controlled everything, the local people began demanding that it either fill in the rubbishy canals or cover them up and use them as sewage pipes. In Suzhou, about a fourth of the rivers have been filled in over the past 50 years and some have ended up as roadways. When the rivers and lakes disappeared, the flood problem got worse.

It seems pitiful to reflect on the fact that these towns were known for their water resources, but the inhabitants were unable to preserve the resources that their ancestors had depended on for generations. The people's response to the problem was just like that of the water-deprived parts of northern China in the quest for clean water—turn to the aquifer and start digging wells. As the size of the towns and cities increased, the demand for water increased so much in just a few years that the underground water supplies were disappearing at an alarming rate.

The cities of Suzhou, Wuxi, and Changzhou ended up with 2,800 water wells in all. The effect of that was to create an 8,000-square-kilometer cone of depression that stretched all the way to Shanghai. On average, the water table has fallen from 1 to 3 meters per year and, in some regions, by much more. The rapid pace of deterioration or depletion of underground water resources obviously cannot continue and will end up threatening both the quantity and quality of water in the entire region.[3]

China has joined the rest of the world in using all sorts of modern technology. But unfortunately, the very water flowing by or into the houses stinks. Can this really be called development?

"BEAUTIFUL WATERS OF LAKE TAI"

The water was so clear all the way down to the bottom of Lake Tai that I could see the fish and crayfish clearly. I was inspired and wrote the song at one stretch.
—Ren Hongju, referring to his popular song "Beautiful Waters of Lake Tai" (1978)

FOR THE CHINESE, THE NUMBER 3 has magical properties because it gives a sense of completeness—like three things bound together or the legs of a tripod. The number comes up whenever the topic of pollution is raised, and the thing that immediately springs to mind is "three rivers" and "three lakes"—meaning the most polluted bodies of water in China.[4]

Topping the list of "three lakes," unfortunately, is the Yangtze delta's Lake Tai. It is China's third largest lake, one that ties together an entire latticework of rivers and rice paddies in a vast "water land."

The area around the lake covers 36,500 square kilometers, touching Jiangsu, Zhejiang, and Anhui provinces and extending to more than thirty counties, cities, and towns, including Shanghai. There are more than twenty major rivers affected by the lake, which, in total, reach for 120,000 kilometers. (*Appendix 11 — Location of Lake Tai*)

Lake Tai was a scenic place for dignified hermits in ancient times. Until 1980, it was quite clean, but when just about every river and stream feeding into it turned into a sewage dump, the mess was visited upon Lake Tai and the results have been virtually irreversible. In effect, Lake Tai has become a massive waste disposal site for industry, agriculture, and the large number of people living near it.

Each year more than 3 million tons of chemical fertilizers and 100 tons of pesticides are used in the lake region. Research has shown that more of these substances are applied than is necessary to achieve maximum output.

In the last twenty years industry has grown and the flat plains have become known for their innovative economics and an explosion in the number of TVEs. In those two decades, the number of industries in the lake region has almost tripled and they have produced enormous amounts of wastewater, about a billion tons annually, much of which ends up in local streams and lakes.

The growing population caused another increase in pollution. That figure of 40 million people in the area comes out to almost a thousand people per square hectare. These people produce most of the waste that ends up in the lake. In 1993 alone there was 450,000 tons of garbage—everything one could imagine: organic matter, styrofoam cups, throwaway chopsticks, plastic bags, cast-off toys, batteries—that ended up in the lake, along with 880,000 tons of animal waste. In all, that amounted to ten times the amount of industrial waste that ended up in the lake.

The barnyard and human waste had large amounts of nitrogen and phosphorous, which led to the growth of large patches of algae, something that was seldom seen in the lake in the 1960s. By the 1980s it was prolific, and by the 1990s almost no part of the lake was without it, so the entire body of water looked like a gigantic flower. That upset the oxygen balance, and the stocks of fish and crayfish were either significantly diminished or were wiped out altogether.

The water quality kept deteriorating and people living in small cities and towns near the lake were being told not to drink water from the tap in some cases. Larger cities such as Wuxi and Jiaxing had to close their water purification plants occasionally and reduce output during the summer algae growth.

And after years of living with high levels of pollution and of organic nutrients in the food chain or tap water, local people began to show significantly high rates of cancer and a number of infectious diseases.[5]

Previously we looked at the policy, applied to the Huai River, of setting a deadline for meeting pollution standards. The same approach was taken in the case of Lake Tai. On New Year's Eve 1999 the central government closed 128 factories to highlight its three-year campaign to clean up the lake. Today, according to officials, nearly all factories meet government pollution standards.

However, a simple trip to the shores of China's third largest lake will disabuse a person of that notion. The government somehow missed its target. The water is a yellowish brown with specks of dark green algae. All along the shoreline there is a paintlike scum that smells rotten. Official tests have shown that the lake water is still among the state's worst in quality ratings.

The failure of that 12-billion-yuan cleanup project prompted many people to take a closer look at it. Experts say that simply closing a few factories was not enough because the farm and household waste was doing even more damage. And some are nostalgic about the good old days of organic farming. During those halcyon days of yore of two decades ago, each village around the lakeside city of Wuxi would dredge up to 20,000 boatloads, or 50,000 tons, of silt for the spring planting. That meant 27 million tons of organic mud being removed every year at that time. But anyone could tell you that it is impossible to tell farmers to go and clean their own shit out of the rivers and lakes these days.

Even if people finally see that the south's water pollution is every bit as bad as the north's water shortages are, it may be a bit too late. That's because most of the rivers and lakes in China's southeastern costal region are so badly polluted. The possible water shortages in the "water world" have threatened the economic development of the region and have forced officials to do something about them. The central government recently approved another 200 billion yuan to be spent over the next ten years on sewage plants and dredging lakes. But increases in the amount of pollution discharged may outstrip the treatment process. As is the case in the north, a better pricing system, one that forces people to save water, seems to be a better way to deal with the pollution problem. Each ton of wastewater can contaminate three tons of clean water.

There is some room left to save water. Factories in the Lake Tai area use four times the amount of water that their U.S. counterparts do for the same output. Each person uses an average of 180 liters of water per day, a higher amount than in many developed countries, and 30 percent of the region's water is lost from leaky pipes before it even gets to the tap.

A change in consumer eating habits is another possibility. Some wealthier Chinese are now looking for cleaner, organically grown food. And even though farmers say they will never go back to dredging mud and feces, the demand for cleaner food may yet make organic farming a lucrative business.

But there is no magic solution to the lake's problem that will take care of everything immediately. Enough chemicals have already found their way to the beds of lakes and rivers to ensure that there will be nitrogen and phosphorous released for at least twenty years, experts predict. They warn that in view of the fact that major rivers on other continents have taken somewhere from 30 to 100 years to clean, Lake Tai could face a long uphill struggle.

HAPPY EVENT OR DISASTER?

Tall trees fringe the skyline in the wilderness,
In the clear water the moon seems so near to me.
        —Meng Haoran (A.D. 689–740), Tang Dynasty, describing Zhejiang province

THE FLAT PLAINS OF CHINA'S EAST COAST end in the southeastern semitropical hills of Zhejiang and Fujian provinces. The coastal area has many rivers, including the Qiantang, the largest in Zhejiang, and the Min, the largest in Fujian. These rivers are constrained by the topography and none is more than 600 kilometers long. The largest river valley is only 60,000 square kilometers in size. But in terms of water quantity, they are not so constrained. The drainage area of the Min River is not even a tenth of that of the Yellow River, but in volume it is actually larger than that of the Yellow.

But the hilly regions here have also gone through a long process of deforestation and degradation.[6] Although 80 to 90 percent of the hills of Zhejiang and Fujian used to be covered with forests, by 1949 the figure in Zhejiang had dropped to 39 percent and in Fujian it was 48 percent. It was an area where rhinos and elephants had roamed in the very distant past, but one that, by 1949, had lost its unique features to ecological destruction, a process that continues to this day.[7]

But even as late as 1949, the area had not completely lost its wilderness, and what primary forests remained still provided enough cover for some 4,000 South China tigers to prowl. A decade later, during the three years of the Great Leap Forward

(1958–60), more than a thousand of these magnificent animals were killed.[8]

The destruction of the forests in the hilly parts of Zhejiang and Fujian continued after the takeover by the Communist Party in 1949.[9] That led to an increase in the amount of soil erosion and the number of mudslides. Many local waterways also became "hanging" rivers, a phenomenon usually associated with the north. The sediment gradually accumulated and moved lower down and put a halt to the river traffic in the waterways. There was a dramatic increase in the number of droughts and floods that hit the hilly regions as well.[10]

But compared to other parts of China, the region still has one of the highest rates of forest cover, for one reason — its proximity to Taiwan. The decades of military hostility between the mainland and its "renegade province" meant that the region was believed to be too vulnerable to support any major construction projects. Because of this, the region was spared the injection of state investment in heavy industry. As a result, the water quality of the rivers of Zhejiang and Fujian is somewhat better than the quality of most other rivers in China.

However, when the government began economic reforms and the process of opening up to the outside world in 1978, the two provinces were put at the head of the development program. Industry and agriculture began developing at a hectic pace in both provinces, and the mountains and rivers took a hit. The local people, who had been known for their cleverness, were eager to demonstrate just how clever they were. Early on they grasped the essence of the now-famous statement, widely attributed to Deng Xiaoping: "to get rich is glorious." To better compete with the more developed cities such as Shanghai, they started developing profitable businesses — at the expense of the environment.

Zhejiang's Yandang Township sits with Yueqing Bay to the east and the Yandang mountains to the west and is a classic tourist region. In the early 1980s, a shiftless but enterprising local farmer found a method to extract miniscule amounts of gold from discarded electronics equipment. That led to buying up all kinds of cast-off electronics goods, computers, and TV parts. He used considerable amounts of sulfuric and nitric acid to treat the equipment and extract small amounts of gold and other valuable metals. In two villages they were able to extract slightly less than a ton of gold, silver, and copper each year. But—and this is a big but—in doing so, they had to process colossal amounts of discarded materials, as much as 920 tons, and use 1,840 tons of nitric acid, 276 tons of sulfuric acid, 6,000 tons of hydrochloric acid, a ton of hydrofluoric acid, and a ton of calcium cyanide.

According to local people, the sulfuric and nitric acid concentration caused such a strong reaction with the cast-off electronics parts that "it looked like a mushroom cloud after an atomic bomb explosion." Not surprisingly, the trees on the surrounding hillsides could not withstand these and other poisonous gases. Large amounts of sulfuric acid residue and coppery liquid generated during the extraction process were dumped carelessly during the night as well and went into local streams, which flowed out of the mountains and into larger streams and rivers. This industrial development made the river water downstream from the two villages no longer potable and affected water drawn from wells.

Local farmers were forced to purchase their water supplies from elsewhere. No one dared grow rice anywhere near the area because it would be too dangerous to eat. The health of people in the area declined, and 80 percent of the local youth could not pass a basic physical examination for military service.

On August 12, 1998, someone discovered seventeen large steel barrels of chemical waste in Pan'an county in a remote part of Zhejiang with contents so toxic that the fumes killed nearby trees and grass. Sitting a few meters from a mountain stream that flowed into Nanjiang Reservoir at Dongyang, they posed a threat to thousands of people in two counties.

On August 15 the Pan'an Public Security Bureau got a special group to investigate the matter, but they were thwarted at every turn. On August 24 another six tons of toxic waste was found near a mountain stream full of dead fish. That stream also flowed down toward a reservoir. The situation looked so bad that a special investigative unit was called into action. They spent a large amount of time and money notifying the public of the danger via television and other media. There was a reward of up to 40,000 yuan for information on the origin of the dangerous materials.

Because of the large number of chemical factories in Zhejiang province, the investigative unit traveled thousands of miles in search of the cause. The culprit was finally apprehended on April 4, 1999. He was a manager of a local chemical plant and offered what he thought was a good reason for his action: "Other plants dilute their waste and dump it directly into local waterways and lakes. So I decided to collect our waste in steel barrels instead, and for that I'm the one who's considered a criminal."

He had a point in a way. In discussing this matter with the provincial environmental bureau, the author was told by one staff member that such dumping of chemical waste into local waterways was indeed a serious matter, but it was also a common practice done in all sorts of underhanded, illegal ways, with little risk of discovery and even less threat of punishment. With more than 3,000 mostly small chemical plants in Zhejiang, many of them the result of the economic reforms, the provincial environmental protection bureau simply could not monitor them all.

The situation in Fujian is not much better. Pollution there is also on the increase. Taiwanese investors came and set up a large number of factories to take advantage of the cheap labor and lax environmental regulations. The cheap products they produced sold well both in China and abroad. But the environment and people's health paid a heavy price.

The Min River was one example. It drains over half the province's territory. In 1981 less than 360 million tons of waste was dumped into the Min annually. By 1990 it was more than 600 million tons.

Another cause of this decline in the health of Fujian's rivers was the more than 120 pulp mills built in the last several years along the upper reaches of the Min, which discharge large amounts of untreated wastewater into it. The water quality has deteriorated overall, but the worst time comes during the dry season, when what is normally water in Shuikou Reservoir becomes nothing more than a brownish black muck.

The hydroelectric power stations built stair-step-like on many rivers in the hillier regions had a negative effect on the ecology because in tampering with nature they caused pollutants to concentrate in a series of reservoirs.[11]

Decades of central government indifference kept the hilly region undeveloped, but at least there was a cleaner environment than in most other coastal regions. The "opening up" gave people a chance to improve their lives, but they turned instantly to polluting their water. A Chinese sage said: "In good fortune there lurks calamity, in calamity there lies good fortune." Someone will study this and clean up the rivers—eventually. Unfortunately we only have a limited amount of time. Pollution must be stopped now.

"IN MOONLIT PEARLS ONE SEES THE TEARS IN A MERMAID'S EYES"

The estuary of the Pearl River is becoming a "desert" where our future generations won't find fish to eat.
—Liang Song, the former head of the State Oceanic Administration's South China Sea Branch (2001)

THE SOUTHERN PROVINCE OF GUANGDONG has a range of mountains on its northern edge that stops the cold air from the north and leaves Guangdong with almost no winter climate. Its numerous streams flowing down from the mountains all move in one direction, toward the Pearl River, the largest river in southern China. The Pearl River is 2,215 kilometers long. It drains approximately 450,000 square kilometers of land and has the West, North, and East rivers in its reach.[12] *(Appendix 12 — Pearl River Drainage Basin)*

Like many of China's coastal areas, Guangdong was generally overlooked by the central government during Mao's time. But the 1978 reforms caused it to join its coastal neighbors in being pushed to the forefront of development precisely because of its peripheral location. First there were the special economic zones (SEZs) of Shenzhen and Zhuhai, followed quickly by Hainan Island, which was first made into an SEZ and later a province. The southern coastal region, with its proximity to Hong Kong, became a powerful magnet for development and investment that began there and later moved inland.

The speed of development was certainly helped by the enormous amounts of water available in the serpentine Pearl River system. Thanks to the largest amount of rainfall in the country, the Pearl River averages 341 billion cubic meters of water a year, just slightly less than the total amount for the considerably larger Yangtze and six times the annual amount for the Yellow. But the breakneck pace of development caused the region to see its share of destruction that had a deleterious impact on water resources.

Let's take a look at what has happened over the last decade or so in the delta region, which has a large network and maze of tributaries covering the only flood plain to speak of in the entire region. It accounts for only 0.48 percent of China's landmass, but fully 13 percent of all the nation's water is there. The people of the region, over the course of several centuries, devised all sorts of production methods, such as fish farming in ponds surrounded by mulberry trees, sugar cane, or fruits, that encouraged the healthy circulation of organic materials in a biological chain. For a long time it was known as a "paradise of fish and rice."

Then along came modern agricultural methods with large amounts of chemical fertilizers, pesticides, and other inorganic materials, and the area's ecobalance was upset. The fertilizers and pesticides joined animal and other wastes in seeping into the waterways, and the water quality in the delta plummeted.

Then export-oriented industries began popping up everywhere over a two-decade period, bringing with them their own pollution. The result has been 50,000 foreign-backed enterprises with more than U.S.$200 billion worth of investment all over the delta, where, often because of ignorance, their untreated wastewater has begun to pose a serious threat to the entire ecology.

In the Shenzhen SEZ, water shortages and pollution have become high-profile problems in the last decade. In 1991, in fact, the water shortage was so serious that most food and beverage, textile, and cloth printing and dyeing factories had to cut their hours of operation or shut down completely.

The nearby city of Huiyang suffered a similar catastrophe when local rivers began filling up with wastes from electroplating plants and cattle hide processors, which affected the lives of 350,000 people. The wastes made their way into the East River, meaning that ultimately they ended up in the Pearl River, seriously affecting water supplies in the lower reaches.

By 1993 the water shortage had reached such an extent that Guangzhou, which sits alongside the Pearl, had a major water deficit. That problem kept reappearing over the course of the next several years. By 1995 the shortfall was 400,000 tons. In 1999, as the city's water resources suffered from more deterioration, the shortage was the worst it had been in fifty years.

The people in Guangzhou were shocked into reality by the lack of adequate clean water in a city that used to have an abundance of the stuff, so they learned to focus on what they began to call the "two 9-percenters." The first 9 percent referred to the annual increase in the city's need for water, which over the past sixteen years has doubled. Its water needs are more than double the average for all of China's major cities, giving Guangzhou the highest demand in the country. The second 9 percent was the annual growth in wastewater in Guangzhou, which on average came to more than 100 million tons a year. At that rate, every year the city would need one more wastewater treatment facility that could handle 300,000 tons a day. Everyone knows how costly such facilities are, even for the Pearl River Delta, where per capita incomes are some of the highest in the country.

At first it appears that treating the wastewater would benefit the lower reaches. But the situation in the delta is not so simple. Guangzhou and all the cities of the delta and along the lower reaches of the West and North rivers are affected by tides that push the sewage discharged along the lower reaches back upstream. That means that even cities such as Hong Kong are considered part of the overall river system of both the upper and lower reaches of the Pearl—and part of the problem. Unfortunately, the region is a highly developed commercial entity where the only concern seems to be making money and the environment is not even a distant second-place concern. A good testament to this lack of concern is the fact that even the intake pipes of many water purification plants are surrounded by large amounts of sewage, garbage, and human feces.

As industrialization spread inland from the Pearl River Delta to the rest of Guangdong province, the pollution followed. In 1994 provincial environmental authorities found that only eighteen of the thirty-three sources of drinking water they tested had drinkable water after purification. From 1990 to 1995 the amount of wastewater discharged into waterways went from 2.5 billion tons to 3.49 billion. Pollution in the Pearl River does 3 billion yuan worth of damage to business every year.

As pollution has gotten worse, it has affected smaller coastal waterways such as the Little East River in the western part of the province, where the pollution levels have been very high. During the dry season, the ratio of polluted to fresh water is 5:1. The 200,000 people living in the city of Wuchuan on the river's lower reaches have suffered along with the paradise, which is becoming more polluted.

In eastern Guangdong, near the city of Chaozhou, the Feng River has had such high levels of pollution that people have had to turn to bottled water not just for drinking but for other uses as well. And in Huizhou, where the East and West branches of the Pearl were the traditional sources of drinking water, pollution has gradually become a problem. It comes from 200 small cottage industries that dump more than 14 million tons of industrial waste and 100 million tons of untreated household waste into the river. There have been reports of colon bacillus levels twenty-four times the national average, and as many as 500 water fleas per liter of river water.

Thanks to the 1.8 billion tons of pollutants discharged into the Pearl River estuary, it is now the second most polluted water along China's 18,000-kilometer coastline.

For years the estuary was famous for its huge prawns and aquatic life. Those are by now nearly extinct and the eight major canneries there have little work—even at the height of the fishing season. As the fish and seafood supplies disappeared, there was an effort to develop more fish farms in the coastal waters. But the sea fought back and wouldn't allow humans to behave so blithely. The water at the edges of the Pearl River Delta has been hit in recent years by increasing numbers of red tides. Many of the would-be fish farms went belly up.

But it was Hong Kong, the "Pearl of the Orient," where people suffered the most. There have been cases in recent years in which people who consumed large amounts of fish, pork, and chicken died from the effects of eating animals raised on various types of contamination.

And speaking of pearls, some say that in terms of beauty those from the Middle East cannot hold a candle to those from Japan, which cannot hold a candle to those from the South China Sea. And they were oh so attractive, those South Sea pearls, but they're now dying a slow death—their quality and quantity are declining as pollution in the estuary gets worse.

"In moonlit pearls one sees the tears in mermaid's eyes," wrote a Tang Dynasty poet. Pearls themselves don't shed tears. Perhaps they should. Sea turtles shed tears, but scientists say that's just to purge their systems of salt. They should shed them for a different reason. And maybe one day humans too will shed real tears for the death of a river system and a sea.

PRESERVING THEIR ROOTS

The river forms a green hazy belt,
The mountains are like hairpins of blue jade.
　　　　　—Han Yu (A.D. 768–824), Tang Dynasty, describing the Li River,
　　　　　a tributary of the Pearl.

IF ONE MOVES WESTWARD FROM GUANGDONG, along the West River (Xijiang), a tributary of the Pearl, one encounters the mountains of Guangxi, or the Guangxi-Zhuang Autonomous Region.

Guangxi is very mountainous, with 64 percent of its area covered with hills and mountains. The forests play an important role in the ecology and the water resources. Like its neighbor to the east (Guangdong), Guangxi had large-scale immigration during the early Qing Dynasty, leading to large land reclamation efforts, especially to provide land to cultivate sugar cane—at the expense of the rich forests. In about a 300-year period, the forest cover had dropped incredibly, from 80 percent to 8.5 percent by 1949. The region is still struggling to get out of this cycle.[13]

The thinned-out forests were generally unable to provide adequate cover on the mountains. The water mingled with the rocky soil, giving what had begun as fairly bleak, barren peaks an even bleaker, more barren look. In 1949 Guangxi had about 10,000 square kilometers of significant water and soil erosion. By the end of the 1980s that had grown to 30,600 square kilometers. Large amounts of nutrients in the soil were leached away by water, and the annual nitrogen loss is now estimated at 103,000 tons. For phosphorous it is 34,000 tons; for calcium, 1.4 million tons; and for organic matter as a whole, 39 million tons. Arable land was lost and farmers were forced to use larger amounts of chemical fertilizers, resulting in increased granulation of the soil and a decrease in the organic matter, which exacerbated the soil erosion problem.

Deforestation helped cause an increase in the serious droughts and floods over the years, and the response was the same as it was elsewhere in China—more water retention and control facilities. But, inevitably, the serious soil erosion caused problems with those facilities in Guangxi, especially with the buildup of sediment in many reservoirs.

The ecological degradation led to climate changes. Historians have written that from 1618 to 1943 major droughts hit the region, on average, once every thirty-three years. But from 1946 to 1972 the interval fell to every six years and in the 1980s to every two years. There were four major droughts in the three-year period from 1989 to 1991.

Then, like many other parts of China, Guangxi was hit with the double whammy: more frequent droughts followed by more frequent floods. In the 1990s the Pearl River valley, downstream in Guangdong, suffered from a series of major floods whose source was traced to the mountains of Guangxi. Now hundred-year floods hit the region every three or four years. Between 1988 and 1998 flooding did 183.7 billion yuan worth of damage in Guangxi and Guangdong.

The Pearl River floods originate in Guangxi, it's true, but the province is also a major source of pollution. According to a 1998 State Environmental Protection Bureau report, which separated the 23 major rivers of Guangxi into 66 sections, there were 33 whose water quality was at level III, 21 with quality at level IV, and 12 at level V or worse. That situation seems to be getting worse: Cyanide levels in the Gui River have risen, and acres of dead fish have been found floating on the surface of the Dragon River (Longjiang). In Tian'e county, water intake from the Hongshui River, which is the Guangxi section of the Pearl, had to be halted because of high pollution levels. The local people have been left to comment: "The Red Water River (Hongshuijiang) has become a 'Black Dragon River' and Guangxi's Diao River is now a 'river of poison.'"

Guangxi does, however, still have some beautiful waters. The old comment that "the mountains and waters of Guilin are the finest under heaven" refers to the magnificent karst formations around Guilin, which have become one of China's major tourist attractions. No question about it, Guilin's scenery along the Li River is something of great beauty and pristine serenity. But the Li, like many of the other rivers of Guangxi, faces an increased threat from pollution that could have economic consequences. It was back in 1976 when Deng Xiaoping, during a tour of Guilin, noticed the increased levels of pollution in the Li and commented: "If, in your drive to increase production, the Li River is polluted and the local environment despoiled, then everything will be for naught!" Three years later, *Enlightenment Daily*, a newspaper that was in the forefront of investigative reporting on China's environment, ran an article entitled "Rescuing Guilin" in which it revealed that pollution was causing the waters of the Li to deteriorate into a turgid black mess, while the rocks and forests on its banks suffered as well. At the same time, there was an increase in the number of low-water periods in the Li, and tourism in the Guilin area began to fall off.

Fortunately, the Li had an international reputation and there was a real effort to rescue Guilin. And the local economic reforms led not to greater industry and increased pollution levels but to an increase in tourism that was so large that locals declared that the Li had become "a treasure trove of gold and silver." The revenues that Guilin's popular sites generated prompted the local government to improve the natural beauty even more by closing down forty-nine plants that were still polluting, an action that had an immediate effect on water quality in the slowly meandering Li. There is still room for improvement, however.

But cities and towns along the Li's banks were guzzling water at a rate of 150 million cubic meters a year and were causing the Li to have periodic dry periods. In 1999 there was some really bad news: A large part of the popular Cat Mountain's forests had been hit by an infestation of insects that had killed more than 50 percent of the trees. People who were familiar with Guilin knew that the affected region was a most important water resource for the upper reaches of the Li. If the insect problem spread, the results would be dreadful for Guilin's scenery.

Now, all of this may have been shocking, but it certainly would have come as no big surprise to anyone who knew about the almost bottomless appetite of the locals in their taste for exotic, rich meals. A vast array of interesting animals, from cobras and boa constrictors to pangolins and monitor lizards right on down to every kind of creepy-crawly critter, as well as a rainbow array of wild birds, could end up as supper on the tables of local restaurants.

During the past five years, several hundred zoologists and other specialists have combed the mountainous slopes of the region to make the first real tally of the remaining amphibians, reptiles, birds, and other members of the 120 species of animals that once roamed the tropical and semitropical forests. The results were disappointing: The number of monkeys, snakes, lizards, turtles, and other inhabitants of the wild land had decreased dramatically. And some of the rare animals—the tigers, leopards, gibbons, and red deer—have just disappeared. The searchers did not find a single pangolin, an animal that used to be very common to the region.

A less exotic but equally destructive Guangxi practice has been the indigenous handicrafts industry of sculptures made out of native wood. Unfortunately, this very destructive art form, which has been taught for a long time, entails not only chopping the trees down but also completely digging up their roots, meaning no second growth and the undoing of any attempts at soil conservation. In the search for materials, the artists have gone to Cat Mountain, the most important water source.

In the days of dense natural tree growth, the native practices were practically harmless. But now the wanton destruction of trees has left a lot to worry about: Where will the future roots, either tree or human, be? The roots need to be preserved now for the sake of the mountains, the clear rivers, and the trees.

CHAPTER SEVEN

SOUTHWESTERN CHINA

"SOUTH OF THE CLOUDS"

The endless forest is our home,
Tigers and wolves are our dogs,
Peacocks and other wild birds are our chickens,
Medicinal herbs in the mountains are our riches.
　　　　　—An ancient folk rhyme of the Hani (Akha, in Laos),
　　　　　　one of the ethnic groups of Yunnan province

SOUTHWESTERN CHINA CONSISTS OF A LARGE LANDMASS from the Yunnan-Guizhou Plateau to the Qinghai-Tibetan Plateau. The region is the source of many of China's major rivers and the headwaters of the great rivers of South and Southeast Asia. Both the Yangtze and Yellow rivers have their headwaters on the Qinghai-Tibetan Plateau, while the origins of the Pearl River can be traced to the Nanpan River on the Yunnan-Guizhou Plateau.

Vietnam's Red River, which originates in Yunnan, and several tributaries of Myanmar's Irrawaddy and Salween rivers and Southeast Asia's Mekong (Lancang in China) have their headwaters in the snowcapped mountains of the Qinghai-Tibetan Plateau. India's Brahmaputra also originates in Tibet as the Yarlung Zangbo, and Pakistan's Indian River starts in Tibet. *(Appendix 13 – Major Rivers Originating in Southwestern China)*

These rivers nurtured the ancient civilizations of East, South, and Southeast Asia and so have been the life source of the world's largest populations. In this sense, it can be said that the southwestern part of China made a major contribution to some of the world's great civilizations.

But the dry periods that hit the Yellow River and the floods that bedevil the Yangtze River can in large part be traced to the headwaters of these two waterways in the southwest as well, and the destruction of the ecosystem there. They are therefore a symbol of the deterioration of water resources that has afflicted the southwestern part of China for some time. As more disasters befell the lower reaches of the rivers and the developed plains much farther down, people there finally began to take a look at something they had ignored for so long — the situation around the headwaters.

If we follow the Pearl River upstream, we first enter the Yunnan-Guizhou Plateau. The water system of Yunnan province (the name means "south of the clouds") is perhaps the most complex in all of China. The Pearl originates in the eastern part of Yunnan, which also contains an upstream portion of the Yangtze (known in the area as the Jinsha, or "golden sands"), the Lancang (Mekong), the Nu (Salween), and the Irrawaddy. These five major river systems have more than 600 tributaries and branches and 221 billion cubic meters of water in Yunnan alone. Per capita annual access to water in this area is 6,297 cubic meters, or 2.6 times the national average.

These water resources support China's richest and greatest habitat in that province, with 18,000 different varieties of plant life, or 60 percent of all the varieties recorded in China. Many of them can be found in the province's large tropical forest reserves.

The trees and plants have helped maintain the rivers' bounty. But people have forgotten or simply ignored the very treasures that surrounded them and for centuries have randomly hacked away at or burned down the forest cover of the mountains as if trees were an enemy of civilization instead of an ally.

From the 1950s to the 1980s Yunnan saw the largest period of destruction in the history of China's forests. More specifically, from 1950 to 1981, average consumption of Yunnan's forest reserves amounted to 27 million cubic meters annually. That figure outstripped new forest growth by 100 percent and caused the province's forests to decrease from 1.4 billion cubic meters in 1950 to only 980 million in 1981. By 1990 only 23.7 percent of Yunnan had forests — this in a province that had at one time been virtually nothing but forests.

The impact of the rapid deforestation was dramatic. For twenty-one of the forty-six years from 1950 to 1996, Yunnan had major floods. Those floods as well as mudslides affected 60 percent of the province's population, 30 percent of its farmland,

and 70 percent of its industrial and agricultural output. The cost? On average, it's been 2 billion yuan annually, but in 1996 it came to 3.2 billion yuan and in 1997 it was 4.5 billion yuan.

Perhaps the greatest sign of just how bad the ecosystem has become is the increasing frequency and severity of mudslides, a phenomenon that now affects every major river system in Yunnan. At the Pearl River's headwaters, for example, there are 68 major ravines caused by mudslides, while at the Red River's headwaters there are 248 and in the Lancang River Valley 666. Around the Nu (Salween), there are more than 500 of them, and they have had an adverse effect on the local people, the more so since the slides often occur with little or no warning. Since 1949 they have taken the lives of 5,000 people, and fifteen county-level cities and thirty-four townships currently live under the threat of more mudslides.[1]

While the floods have come more frequently and with greater severity, the opposite effect, drought, has come to be felt more as well. More than 100 counties have been hit in recent years with major droughts, the effects of which were felt substantially in the 300,000-ton reduction in annual grain output.

In spite of the blatant nature of these catastrophes, the people seem to have remained virtually oblivious to the reality and to any thoughts about devising ways to protect the trees. In early 1992 a group of people who visited the headwaters of three major rivers in Yunnan in search of azaleas found the Yunnan yew trees (*hongdoushan* in Chinese, *Taxus yunnanensis*) instead and left with some samples of its bark, in which they were interested. That caused a surge in interest in the tree for its commercial purposes, and over the next two years, the local people, believing themselves to be in possession of a new money source, trekked up into the mountains, chopped down the trees for their bark, and then sold it for a mere 3–5 yuan per kilo.

Later, studies by the American Cancer Society of a chemical compound extracted from the tree, taxol, found that it had been successful in treating certain forms of cancer, especially leukemia. A kilogram of taxol was estimated to be worth $8 million on the international market.

By the time Chinese scientists finally awakened to the fact that the roots, branches, and bark of the yew were of such great value, it was too late—the folks in the mountains of Yunnan had stripped the bark off any yews they could get their hands on and had sold it at bargain basement prices.

Other parts of the forests suffered a similar fate during the period of economic development. Under one policy that encouraged what was known as "compensated transfer," people rushed to sign planting contracts, and in no time at all trees were cut down and replaced by fields of sugar cane, bananas, or pineapples. In the five years from 1992 to 1997, Yunnan's forest cover decreased by more than 70,000 hectares, causing an increase in soil erosion, which now affects 38 percent of all the province's territory.

Yunnan also has tropical rain forests, especially around the popular tourist region of Xishuangbanna, on the southern border with Laos and Myanmar. The area has only 0.2 percent of China's landmass, but within that small space there are more than 4,000 species of trees and plants, or a sixth of the total in China. And they have a multitude of possible uses, not the least of which is medicinal. The area also abounds in animal life, with 102 different species of mammals and more than 420 types of birds.

But even this wilderness has been marred by destruction, first during the rubber-harvesting campaign of the 1950s, then during the period of sending youth down to the countryside after the Cultural Revolution to reclaim thousands of hectares of land. The results of that awful period were a decline in the amount of rain forest from 80 percent of the region in 1949 to 30 percent today. And the destruction has continued. Each year 2 percent of the tropical rain forest is lost to various forms of development, and today Yunnan has eighty of China's endangered animal and plant species.

These changes have also had an effect on the amount of rainfall in the rainforests. According to data from meteorologists, the average annual rainfall dropped from 1,221 millimeters in the 1960s to 1,189 millimeters in the 1970s. At the same time the amount of evaporation increased from 1,394 millimeters to 1,600 millimeters, pointing to a net deficit in the region's moisture on average.

The central government's logging ban that followed the great floods of 1998 was also imposed in Yunnan. Unfortunately for the Yunnan yew, not to mention the hornbill, gibbons, sloth, wild buffalo, and Asian elephants, it was too late.

BRIGHT PEARLS ON THE PLATEAU

The massive pollution control efforts of the past five years have only made us realize how tough and complicated it is to clean up a polluted lake. In addition to resolve and confidence, we need more patience.
—Chen Xunru, vice governor of Yunnan province, 2001

THE PLAINS AROUND THE LAKES of the mountainous Yunnan-Guizhou Plateau have provided a habitat for humans and flora and fauna, but the lakes themselves have suffered ecological damage, again because of the destructive impulses of humans, especially in recent years.

Dian Lake (Dianchi), which lies on the southwestern edge of Kunming, Yunnan's capital, is the most important body of fresh water on the plateau. The lake is a member of the largest group of lake basins in the eastern part of Yunnan. Dian was the classical name for Yunnan (harking back to the Dian kingdom of the fifth century B.C.) and was used to describe a "vast, deep body of water that at times appears to be flowing backward."

Well, unfortunately, that lake is now a pale shadow of its former self thanks to the large-scale destruction of the forests in its watershed, which caused the lake to become shallow as sediment and organic matter were washed into it.[2]

As if that were not enough, the shrinking lake was the recipient of increased amounts of wastewater and toxins. There was no way that this lake would somehow remain clean and free from the ill effects of the modern world.[3] As more pollution entered the lake, the northern end, the one that touches the city, began accumulating a 10-centimeter-thick layer of muck that included 9,700 tons of nitrogen and 4,650 tons of phosphorous. These accumulated in the lake's mud and were slowly released into the lake, making any pollution control efforts exceedingly difficult.

But when Kunming was chosen to host the World Horticulture Exposition in 1999, the largest international event Yunnan had ever held, the provincial government decided to put on a happy face and spend 5 billion yuan on cleaning up the lake. In October 1999, after a brief but thorough cleanup of the water, the lake was unexpectedly hit with a sudden outbreak of blue-green algae so thick that boats were unable to come ashore. The algae then washed ashore and collected in parks, on roads, and on lawns, with some blotches as much as 2 meters long.

People were tempted to say that Dian Lake was suffering from some kind of "ecological cancer," one for which there was no effective cure. Yunnan announced that it would spend another 9 billion yuan to deal with the problem. But people were becoming dubious or even pessimistic about the results. Some experts proposed that Yangtze River water be brought from far away to clean the lake, because flowing water was the only thing that could rinse away the scum.

Yunnan's second largest freshwater lake is Erhai. It is also China's seventh-largest lake. In addition to the usual deterioration, Erhai has had another curious problem, thanks to the impact in recent years of the of the whitebait fish, which was introduced into its waters from Lake Tai in eastern China. The reason for moving the fish seemed logical enough. They were introduced as an experiment into Erhai in 1991 because increased amounts of pollution in Lake Tai had caused a reduction in the size of whitebait catches there. Initially everything seemed to be going very well and whitebait catches in Erhai were running to 530 tons. But just as people were celebrating the successful transplanting of this remarkable fish, they began to notice a decline in the water quality in the lake. Then in September 1996 it was hit by a major outbreak of blue-green algae.

This was certainly curious. How could such a small fish like the whitebait cause so much trouble? According to Du Baohan, of the Environmental Research Institute of the town of Dali, it was the result of the whitebait having disturbed the lake's ecological balance. The whitebait fed off various smaller organisms, which, when they were completely consumed, no longer fed off the blue-green algae. The whitebait had caused a "biological invasion," a phenomenon that got a lot of attention from environmental organizations around the world. The whitebait also consumed the eggs of other fish species and in turn discharged a sort of toxic substance that killed certain fish. The whitebait also reproduced rapidly, and without any predators in the lake to eat them, they quickly moved into areas occupied by other fish that were native to the lake and eventually wiped them out.

This failure was the result of an effort to get short-term economic gains by introducing a non-native species. Nature's balance was created by nature itself and humans should know better than to upset the balance just for personal benefit. Unfortunately, the Erhai whitebait disaster was repeated elsewhere. Lake Dian, for example, has seen the number of its fish species drop from twenty-five to six as a result of similar tampering. And Lake Lugu, in the northeastern part of Yunnan, on the border with Sichuan province, in spite of being relatively remote, has all but lost its famous thin-scaled fish (*xilin yu*), again because of the same kind of experiment with nonindigenous fish species. Across the high lake plateau of Yunnan, studies have found that out of the several hundred varieties of fish known to have existed, two-thirds have virtually disappeared and one-third are so thinned out that the stocks provide little in the way of viable catches.

The effects of tampering with the environment are also apparent in Chenghai Lake, where visions of profits completely backfired. Chenghai is one of only three lakes in the world with naturally occurring spiral algae, but the balance was permanently altered when a decision was made a few years ago to turn the algae into a food product that was thought to be beneficial in treating certain diseases. With provincial government backing, everything was tried to increase production of this valuable algae in eighteen pens in the lake. Unfortunately, the process of increasing production of the algae also produced an increase in wastewater, so by 1997 there were 1.2 million tons of discharge. By the year 2000 the discharge was estimated at 3 million tons, an amount that will certainly undermine the quality of the lake's water.

A similar fate also befell Caohai (the name means "grass sea") in nearby Guizhou province. Caohai is only 45 square kilometers in size but is Guizhou's largest lake and a wetland that is home to 110 varieties of birds, which number around 100,000 in all. But another crucial function of the lake has been its role as a giant humidifier that releases 120 million kilograms of moisture.

Unfortunately, Caohai also became the target of land reclamation for agriculture back in the 1960s. By the 1970s it had shrunk considerably. That not only reduced the lake's ability to act as a humidifier but also caused a substantial drop in the number of black-necked cranes frequenting the area. In spite of such warning signs, the local people were determined to cultivate

Caohai. But there were still obstacles to development, not the least of which was the fact that the lake bottom, when exposed, turned out to consist mostly of rock. Not only was there was a significant drop in the amount of fish caught, but much of the time invested in reclamation came to naught.

In a pleasant reversal of the normal course of events, efforts to develop the lake were all but dropped in the 1980s and the wetlands began to make a comeback, luring back some of the birds and replenishing the fish stocks. The lake is now under international protection, but it has not yet escaped the cycle of population increases, worsening poverty, and degraded ecology.

The people on the plateau at some point will have to reach a compromise with the lakes if they don't want to lose all of their unique flora and fauna.

THE LAST SHANGRI-LA

The azure sky is a basin made of turquoise,
The bright sunshine is our gold jewelry.
　　　　　—Tibetan folk rhyme

QINGHAI PROVINCE AND TIBET have been the destinations of a growing number of tourists because of the magnificent scenery that stays with everyone who has visited. Some sights are only for being seen from afar because they are too high, too cold, and the air is too thin.

Many parts of the Qinghai-Tibetan Plateau have few human inhabitants or none at all, and many people think of it as a kind of forbidden zone. That is far from the truth. In addition to the high mountains and valleys with their dense forest cover in the southeastern part of the plateau, there are the hinterlands that seemed so alien but were in fact a sort of Shangri-La for many plant and animal species.

Qinghai province alone has all manner of wild animals, 103 species in fact, along with 292 bird species and more than 3,000 varieties of plants. In Tibet, because there are so many mature forests that have yet to be felled, there are 2.1 billion square meters of timber. And Tibet has the largest number of wild animals in China.

The region's vast water resources support all this. Qinghai has the headwaters of the Yellow, the Yangtze, and the Lancang (Mekong) rivers, making it the fountainhead of China. In Tibet, which accounts for a major part of the plateau, there are 1,500 lakes, 600 of which are more than a square kilometer in size. The lakes of Tibet account for a third of the lake surface in China.

The northern part of the Qinghai-Tibetan Plateau is known in Tibetan as the Yangtang (in Mandarin, Qiangtang). It is almost entirely without human habitation and is the second most desolate spot on earth, after Antarctica. It runs for more than 1,300 kilometers from west to east and is 4,500 meters above sea level on average. It is treeless but there are a large number of wild animals that inhabit its barren valleys. The heart and soul of the Yangtang are the more than 1,000 separate lakes, most of which are brackish and which total more than 30,000 square kilometers in size. These include Nam Co and Siling Co, which are more than 1,000 square kilometers in size. (*Appendix 15 – Major Lakes in the Qinghai Tibet Plateau*)

By the time people got around to learning of the very existence of these lakes and understanding their importance to the fragile ecological balance, they discovered that virtually all of them had been shrinking. Some had even disappeared.

The finger of blame was pointed everywhere and was in fact a matter of debate. Most scientists believed that it had something to do with global warming. But there were those who blamed the immense deforestation in western Sichuan and Yunnan and said that it had reduced the amount of moisture available to the plateau via global currents. Others held the people who had wantonly hunted down the animals of the area responsible, and said that once the antelope, a favorite target, had been killed off, the wolves followed. That left the rabbit population to grow out of control, burrowing holes everywhere and eating up the grassland. And when the hawks and lynx of the region were killed off, it robbed the area of valuable predators that fed on rats, whose numbers then grew astronomically. The rats, like the rabbits, dug holes everywhere, upsetting the fragile grassland ecology. Or so the theory went. A study of the remote Hoh Xil region found 116 rat holes in just a 100-square-meter area. This "rat disaster" was so out of control that over the last ten years 15 million hectares of grassland on the plateau suffered grievous damage and a third of it turned into nothing but mounds of black earth.

The global warming part of the problem is not something the Chinese can stop by themselves. What they can do, however, and should do, is to take a look at what else has gone wrong and to take steps to correct the mistakes.

One example might be Qinghai Lake (Koko Nur in Mongolian), in the northeastern part of the province, which is so vast that during imperial times it was considered a sea. Its 4,538 square kilometers make Qinghai the largest lake in China. Throughout history it has been fed by seventy rivers, large and small, that were themselves fed by melting glaciers.

Although it has shrunk naturally over the centuries because of a rising Qinghai-Tibet Plateau, it was humans who apparently accelerated the shrinkage. From 748 to 1936, a 1,188-year period, the water level of Qinghai dropped 4.5 centimeters a year on average, creating several small islets. But since 1959 the annual drop in the water level has accelerated to 10 centimeters, a drop of around 3 meters in all.

The culprit is, once again, agriculture and overgrazing. Prior to 1949, the lake valley had practically no agriculture or cultivated land to speak of, only a mere 697 hectares, mostly in barley. But by the end of 1957 that had grown to 4,428 hectares. Then in 1958, during the madness of the Great Leap Forward, hordes of people rushed to the lake area to reclaim land without any guidelines at all, and by 1960 the total area under cultivation came to nearly 49,000 hectares.

There were, however, a considerable number of natural limits to what Qinghai could support in either crops or herds (the latter, too, have grown substantially since the introduction of economic reforms), and many areas under cultivation yielded little or nothing at all. There was also overgrazing, which began an inexorable process, from the 1950s on, that resulted in leaving 4.5 million square hectares of land in eight counties completely barren. Over the past thirty years 67 percent of the land has developed some form of desertification.

Irrigation and overgrazing have drawn the water down quite a lot, so that the seventy rivers that fed Qinghai in the 1950s now amount to only four and the overall flow is down 60 percent. Much of the water never even makes it to the lake.

Of course, by dint of the hard work of reclaiming land and developing agriculture, Qinghai is now self-sufficient in grain production, something of which the Chinese can be proud. But this success came at the expense of the region's biggest ecoparadise. Was it worth it?

The economic expansion went as far as Tibet's doorstep, and the changes it wrought in the ecology are quite disturbing. The high grasslands of Ali, one of the remotest spots on earth, contain the headwaters of the Sênggê Zangbo (*Shiquan*, or "out of the lion's mouth") and the Langqên Kanbab (*Xiangquan*, or "out of the elephant's mouth"), the sources of India's Indus and Sutlej rivers, respectively. Before they completely leave Tibet, these two rivers drain an area that was known for years as a Shangri-La of wilderness. Prior to 1966, both the rivers had plenty of water and grasses grew everywhere, as did red willows and a wide variety of other plants, in addition to cultivated crops such as highland barley, spinach, cabbage, and turnips.

But the numbers of people and animals increased. They congregated around Shiquan township, and their demand for fuel during the cold winter months outgrew the traditional supplies of cow dung. Out of desperation, the people began relying increasingly on bushes and other undergrowth until finally, by the end of the 1970s, the red willows and cattails had completely disappeared and had been replaced by sand. The area now looks immensely bleak and inhospitable, especially during the winter months, when winds whip up the sand and blast the faces of buildings and other facilities. The sand, wind, and sun also threaten the health of humans, with 80 percent of the people of the area suffering from cataracts, pulmonary infections, and frequent nosebleeds.

The Yarlung Zangbo is the largest of the Tibetan plateau's many rivers. The bed of this 2,000-kilometer-long river is, on average, 3,000 meters above sea level, making it the world's highest river. It is third in water volume only to the Yangtze and the Pearl and provides water for the land around both its banks. Its valley has traditionally been a center of politics, economy, and culture. But because of a long history of dense population and destruction of the habitat, the river valley has suffered from significant desertification. There are five counties along the river that put 15 million tons of soil into the river each year, seriously threatening animal husbandry and agricultural development. Sand dunes now meander along the river valley, shifted by the interminable winds.

The lower reaches of the Yarlung Zangbo are below the town of Mainling. Once the river crosses the eastern section of the Himalayas it makes a great horseshoe bend. A group of scientists and journalists four years ago chose that exact spot to begin a trek of the Yarlung Zangbo valley, "the world's last and best-kept secret," from October 29 to December 3, 1998. Along both banks the mountains rose more than 7,000 meters above sea level, and the river itself had a drop of 2,300 meters. It cut through the mountains for 650 kilometers and rushed toward the Bay of Bengal through largely uninhabited areas where there were four major groups of waterfalls, a rarity for such a large river.

But that was several years ago. Now the once pristine Yarlung Zangbo valley attracts tourists and a significant amount of garbage. And the slaughter of many endangered species of animals in the valley by local people goes on unabated. On November 3, 1998, seven black bears were killed. In just a few years' time, many of the wild animals in the valley have been thinned out.

According to some Tibetan legends, that mysterious bend in the Yarlun Zangbo River is the real Shangbala, or Shangri-La. In Tibetan Buddhism it is a mysterious land with no poverty, hatred, hypocrisy, cheating, or death. Many people have spent years wandering among the Tibetan mountains in search of the perfect land.

But every time we conqueror yet another peak or cross another valley, Shangbala still seems to be just beyond the next mountain. It's only when we turn around and look back with just the right degree of reverence at the many snowcapped peaks and valleys we've crossed that we can understand that the very place where we can settle in peace may be the one we were just in.

NOTES

CHAPTER ONE

1. This was no ordinary period, coinciding as it did with the Great Leap Forward (1958–60), which quickly turned into a major disaster for the entire country as man-made and natural calamities colluded to deal the nation a nasty blow. While the specter of famine hovered over the Chinese people and millions of people starved to death, Sino-Soviet relations evaporated. At their worst they produced skirmishes along the immense Sino-Soviet border.

2. The biggest lesson the Chinese leaders drew from the Three Gate Gorge experience was that dams on the Yellow River and other murky rivers needed to follow a policy of "impounding clear water during the dry season and flushing out murky water during the flood season" (*xuqing paihun*). While this might have worked on one dam, when an entire series of dams was built on the Yellow River and when every one of them tried to hold back the clear water and discharge the murky water, less water and more sediment flowed to the lower reaches, causing a general deterioration there. During the flood season, after floodwaters in many reservoirs on the upper reaches (like Longyang in Qinghai province) subsided, less floodwater flowed into the Three Gate Gorge reservoir, leaving even less water for downstream. And in spite of a policy of "impounding clear water and flushing out the murky," the amount of sedimentation washing into the reservoir did not diminish. In fact, the amount of sediment per cubic meter of water actually increased. So a smaller amount of water could carry a larger amount of sediment to the lower reaches. That only made the sediment accumulation problem worse. During the drier season, the emphasis placed on electricity at the Three Gate Gorge dam left relatively little water for the lower reaches — in fact, water flow was barely enough to rinse the riverbed on the lower reaches, let alone flush its dirt out to the sea.

3. The changes in the dry periods have been especially disconcerting. In the 1970s and '80s they mainly occurred in the middle of the northern Chinese summer drought season, in May and June. But by the 1990s there were significant dry periods at times when there is usually runoff—in February, March, and April—and then again the following summer and autumn.

4. In the 1970s the longest period was 21 days; in the 1980s it was 36 days. In the 1990s there were a series of record dry periods: 16 days in 1991, 82 days in 1992, 61 days in 1993, 75 days in 1994, 118 days in 1995, and 133 days in 1996.

5. The grains of sandy earth average about 5 millimeters in diameter and are the primary source of deserts and of little use to agriculture.

6. The river valley below Dongbatou now has 1,987 square kilometers of sandy beach. When the river dries up, the sediment quickly becomes sand and desertification becomes a threat. Farther downstream, the higher elevations along the Yellow River near Heze in Shandong province in recent years have developed 135 square kilometers of shifting sand dunes, a clear sign of advanced desertification.

7. Water shortages have become an especially serious problem on the eastern and northern Henan plains, where per capita consumption is only 270 cubic meters even though 52 percent of the entire province is arable land.

8. The city of Zhengzhou is one example. Because of a plentiful supply of water from the Yellow River in the past it was able to prosper, with smaller rivers such as the Jialu, Xiong'er, and Jinshui also contributing to agriculture and industry. But in recent years, in addition to increasingly serious pollution problems, Zhengzhou has encountered something it never thought possible: a progressive drop in water levels and the drying up of the three small rivers that were the source of its prosperity. Since the 1970s Zhengzhou has depended completely on the Yellow River for its water.

The drying up of the Yellow River has not reached Zhengzhou yet, but the continued decrease in the amount of water around the city has meant that the river is a mere stream whose volume cannot meet the city's large-scale irrigation and industrial needs. There are huge water pipes that were laid for the sole purpose of drawing water out of the main channel that now jut out of the embankment above the dry riverbed for long periods of time. The city suffers from an acute lack of water, with the shortfall sometimes estimated at 600,000 cubic meters a day.

Problems exist elsewhere in Henan, in the ancient city of Luoyang, where water shortages have become quite common largely because the nearby Luo River has been dry for some time now. To make up for the water shortfall, Henan has increased the use of underground water. Subterranean water now accounts for 72 percent of the water the province uses. Henan now uses 9 billion cubic meters of this water annually.

This heavy reliance on underground sources has not come without a price. The eastern and northern plains of Henan have become the sites of cones of depression (*loudou*). The formation of funnel-shaped depressions has also been a problem in the cities of Zhengzhou, Kaifeng, Jiaozuo, and Fuyang. These cities have an additional problem with pollution entering the aquifers. In Kaifeng 52 percent of the underground water has high levels of pollution; in Jiaozuo it's 45 percent. Unless something drastic is done to change water use in the near future, the underground water supplies will go the way of the Yellow River. In Zhengzhou, funnels occur on a 162.2-square-kilometer area. Subterranean water levels are about 62 meters below the surface and falling by 1 or 2 meters annually. There is virtually no moisture at all near the surface.

9. In the summer of 1994 the drought was so severe that twelve of the province's fourteen major rivers dried up and more than 4,000 medium-sized and small rivers suffered from the same problem. In addition, 27,000 reservoirs and ponds dried up altogether and 20,600 water wells ran dry. That left more than a million hectares of farmland without water. Crops mostly failed on 100,000 hectares, and 56,700 hectares yielded no harvest at all.

10. On the Shandong Peninsula, an area that traditionally had little need for water from the Yellow River, large-scale development since the 1980s has led to major reductions in water supplies. The result has been a cut in production at 229 factories in the cities of Yantai, Longkou, Zhaoyuan, and Penglai, with losses estimated at 869 million yuan. The same applies to the city of Qingdao, where the situation is so serious that water is available to residents and industry for only a few hours a day. In 1989 a drought hit eastern Shandong province, and water shortages in more than thirty cities led to 9 billion yuan worth of industrial losses. Qingdao, which has become famous for Tsingtao Beer, suffered losses worth 2 billion yuan.

11. As a result of the water shortages, annual crop losses in Henan and Shandong have amounted to a billion jin (1.1 billion pounds). In 1995 losses were worth 4.2 billion yuan. In 1997 there was the longest dry period on record, and the results were catastrophic. Crops on 133,000 hectares of land withered and died, resulting in losses of 2.75 billion kilograms of grain and 50 million kilograms of cotton.

12. In 1993, total arable land was 570,000 hectares, supporting a population of 5.1 million, with industrial output figured at 30.5 billion yuan.

13. The averages were 1.32 billion tons in the 1950s, 1.9 billion tons in the 1960s, 900 million tons in the 1970s, 640 million in the 1980s, and by the 1990s only 400 million tons.

14. Dongying has proposed building water storage facilities that can hold 1 billion cubic meters by the year 2002. The city has managed to increase its water storage capacity by 50 million cubic meters annually by trying just about every conceivable approach. In 1998 it spent 120 million yuan on forty-six large and medium-sized reservoirs that increased water storage capacity by 120 million cubic meters.

15. In June of 1995, after a dry spell of forty days, water from the upper reaches finally began flowing into the Yellow River near the city of Jinan in Shandong province. The water that arrived was neither clear nor particularly yellow; instead it was a black, viscous slime punctuated with white foam and dead fish on the surface.

16. In Guangrao 84.6 percent of the local people suffer from fluorine-related tooth and bone ailments. In Lijin 13.3 percent have the same problems.

17. On 153,000 square kilometer of wetlands there are more than 800 different species of aquatic life, including the rare Wenchang fish, the Yellow River dolphin, and the Songjiang perch. The area is also known for its enormous variety of wild plants, including the wild bean (*yesheng dadou*) and a number of rare herbs. At one time or another 187 varieties of birds have been spotted on the wetlands, including the red-headed and white-headed crane, the white crane, and the golden eagle. The area is also home to many commercially attractive marine animals, such as shrimp and crabs, and many different types of fish, which in turn have attracted an army of wild birds, including wading birds, swans, and wild ducks.

18. This systematic pattern of ecological destruction has affected the county's once rich water resources. In less than twenty years, 3,000 of the more than 4,000 lakes in the county have completely dried up, and water levels in those remaining have dropped 2 meters on average. Hundreds of rivers have dried up, and of those remaining, nearly all are seasonal. There are only three major water wells available in the entire county, because a falling water table has made other wells worthless, leaving the county with no choice but to dig new ones. In the spring of 1999 some 38 percent of the herdsmen had to leave their homes because of grassland degradation.

19. The hydrometric station in Maduo county reported that the Yellow River flow rate at the headwaters had gone from 30–40 cubic meters per second down to 10 cubic meters per second. In the rest of the province, from 1988 to 1998 water flow was down nearly a quarter, or 25 billion cubic meters.

20. Sichuan's Norgai Marshland used to provide 8 percent of the water resources of the Yellow River, but now it contributes a tenth of that.

21. Shifting sand dunes created by storms are closing in on the dam by as little as 10 or as much as 70 meters a year. In one year the rate was 81 meters. As a result of this deterioration of the grasslands above the reservoir, the Yellow River and its tributaries deposit almost 9 million cubic meters of sediment behind the dam annually. An additional 21 million cubic meters of sediment flow into the reservoir from the increasingly frequent mud and rock slides on nearby mountains, which now directly threaten the structural integrity of the Longyang Gorge Reservoir. A severe mudslide in 1997 almost destroyed the power generators.

22. The Jing River's source is in eastern Gansu. As early as the Qin and the Han dynasties it had become murky. The Malian River, a tributary of the Jing, was said to be filled with "muddy water" at about the same

time. Centuries later, during the Tang and Song, deforestation was increasingly serious in the Ziwu mountains of eastern Gansu, the Tao River valley of the central steppe, and in the mountainous region of southern Gansu. This process continued until the late Ming Dynasty and on into the Qing, as trees were cut in great quantities, even though the province's forested area had already been drastically reduced. What data there is indicates that as much as 10 million cubic meters of trees were felled annually.

23. Statistics show 127 square kilometers of significant soil erosion in Gansu, or 87.5 percent of the Yellow River basin in the province.

24. From 1950 to 1990 grain production remained at the 1950s level in most cases, and for fifteen of those years it actually dropped below that level. Per capita grain consumption also suffered a drop, from 270 kilograms in 1950 to 254 in 1990.

25. From 1996 to 1998 Gansu used 65 million yuan to assist work on more than a million of these catch basins and cisterns, and the amount of irrigated land increased by 1.3 million hectares. Studies showed that when annual rainfall reached 300 millimeters, the average cistern could accumulate 30 cubic meters of water.

26. This include dams in the Liujia, Bapan, Yanguo, and Hongshan gorges, where reservoirs were used as an important irrigation source and a way of turning the drought-stricken land into a major agricultural base.

27. Five years later, in 1974, pumps were used to bring 148 million cubic meters of water from the Yellow River up through a thirteen-tier pumping system to Jingtai, where 20,000 square hectares of barren land were turned into fertile soil. The land blossomed so richly that people who had abandoned it years earlier returned, and the site served as a place to relocate tens of thousands of people. In 1984 the second phase of the Jingtai project began, with the high-lift pump capacity being raised to 266 million cubic meters of water. Ten years later, more than 34,000 square hectares of new land could be irrigated and over 200,000 people returned.

28. Huangzhiyuan, the largest intact plain in Guansu's Yellow River drainage, was once known as the "grain belt of eastern Gansu," but it lost 40 percent of its land to erosion over the past fifty years.

29. By the time of the Yangshao culture (Neolithic), primitive irrigation channels were already being used in a formative agricultural stage in China. By the Han Dynasty, during the reign of Emperor Hanwu (140 to 87 B.C.), there was such a large network of irrigation channels that the emperor went to inspect the remote area of Ningxia on six separate occasions and ordered people from other areas of the empire to move there. Under the watchful eye of newly appointed officials, irrigation channels were built throughout the area, and those that had been built during the Qin Dynasty were restored.

During the eleventh century, another culture emerged in the area. The Dangxiang tribe set up the Western Xia Dynasty, with its capital at Xingqingfu (today's Yinchuan, the capital of Ningxia). Like their predecessors, they inherited a prosperous realm, with cultural and agricultural wealth bestowed by the soil and river water.

A succession of various dynasties, including the Yuan, Ming, and Qing, put a considerable amount of effort into expanding the area and extending the irrigation network. Eventually the Yinchuan Plains had the most complete and well-maintained system of irrigation in the entire empire. Three great water diversion projects carved out during the Qing—the Daqing, the Huinong, and the Changrun—added 7,000 hectares of irrigated area.

30. From 1986 to 1990 Henan and Shandong provinces, on the lower reaches, drew more than 12 billion cubic meters of water annually. In 1989 the figure was 15.5 billion cubic meters, but according to a national plan they were allocated only 12.5 billion.

31. Dengkou county, at the edge of the Great North Bend, is one example, a sandy windswept area lying between the Helan and Wolf mountain ranges, 150 kilometers long. To the west, Inner Mongolia's Wulan Buhe Desert creeps inexorably toward the Hetao Plains at a rate of 8 to 10 meters a year, devouring 66.7 hectares of fertile land en route. Although the desert sands have not encroached upon the Yellow River directly, the area's sharp winds dump as much as 60 million tons of sand into it annually, adding to the already heavy load of sediment that has caused the riverbed to rise so much that near the county seat it is already 4 to 6 meters above the surrounding land.

In the rest of Ningxi the ecological destruction wrought on the fragile land has been horrendous. In the 1980s it took just a few years for an insect infestation to affect more than 90 percent of the municipal areas and counties of Ningxia. More than 80 million trees were cut down in a last-ditch effort, after spraying and other attempts had failed. That was at least half of the province's entire forestry reserves. And, once again, Ningxia was not alone in this disaster. People of Gansu, Inner Mongolia, Shanxi, and Shaanxi had to join the forest destruction to fight the insect invasion that, in the end, robbed the land of its greatest protection against ecological degradation.

32. Some of dams that continue to take a toll on the river are the Wanjiazhai water diversion project, the second phase of the Jingtai Dam, the Daliushu, and various projects to provide relief to the Qinwangchuan highland. Problematic water diversion projects include that to the Xida River, a tributary of the Shiyang River; to the Hei and Huang rivers; to Qinghai and Nansi lakes; to the Petroleum and Chemical Complex in Shandong; to the cities of Jinan, Qingdao, and Yantai; and to the Wei Mountain irrigation project.

33. In 1997, while several grandiose diversion projects were being built on the middle and upper reaches at a cost of several billion yuan, it came as no great surprise that on the lower reaches, Shandong had one of the longest dry periods—226 days—which brought losses valued at 13.5 billion yuan.

34. Deshengxi township, in the Zhunge'er banner (similar to a county) of Yikeshao league (administrative area), loses 18,800 tons of soil for every square kilometer of land. That figure runs as high as 60,000 tons in some places.

35. Of the 170 million tons of sediment that flow into the Yellow River from Dongsheng, 85 percent is coarse sand.

36. The "'middle reaches'" is generally considered to be the area from Hekou Township in northern Shanxi to Mengjin Township in Henan province to the south and east of Shanxi. Precipitation in this area is quite good, with average annual rainfall of from 400 to 800 millimeters, in addition, there is a greater amount of rain in the Luliang, Qin, and Taihang mountains.

37. The Yan dumps 258 million tons of sediment a year into the Yellow River. Although this has been reduced in recent times to 220 million tons, showing some real improvements in soil conservation, it also is simply the result of less rainfall and lower water levels.

38. The Luo River is the source of an estimated 200 million tons of sediment a year, while the Jing is even more silt-laden. This is because its headwaters are in eastern Gansu, where, as was noted earlier, the desertification problem is getting more severe. On the upper reaches of the Jing there is the Malian River, which, since the Qin Dynasty, has never been anything more than a muddy river with 196 kilograms of sediment per cubic meter on average, the highest of any of the hundreds of rivers feeding the Yellow. Today, it is estimated that 300 million tons of sediment are washed into the Yellow from the Jing annually.

39. Taibai mountain, 3,767 meters high, is perhaps just as important a source of water for Xi'an as are the nearby Qin mountains. It lies at the juncture of Zhouzhi, Taibai, and Mei counties, an area that became the primary target of developers in the early 1990s, with officials from all three counties pushing various plans that ended up seriously undermining the ecology. The result was the destruction of the protective plant cover and a reduction of its ability to maintain water resources.

40. There are now small droughts on an annual basis. Every three years it is hit by a medium drought and every ten years by a serious one. As if that weren't grim enough, there are the monsters, the catastrophic droughts that hit about every thirty years or so and last for two or three years. Since 1949 major droughts have affected up to 1.5 million hectares of land in the province and have had dire human and economic consequences. That situation has actually worsened in recent years. In

1994 drought destroyed 2 million hectares of crops and more than 3 million people had no access to drinking water. The cost of water from the Wei River rose to 10 yuan for 30 kilograms, and a major part of the grain crop was destroyed.

41. It is bounded on the western side by the Luliang mountains and on the east by the Taihang range. The Yellow River forms Shanxi's southern border. To the north of the province is the Great Wall and Inner Mongolia.

42. Even before 1988, what few woods remained were set upon systematically, and 40,000 cubic meters or more of timber were chopped down annually.

43. The reservoir started impounding water in 1961. Since then, the amount of sediment that has accumulated has reduced its capacity by about 300 million cubic meters. This has seriously affected flood control, water supplies, and irrigation for more than 100,000 hectares of fertile land in and around Taiyuan.

44. One result is that since 1979, the amount of underground water that naturally replenished the Fen has been reduced by 85 percent. This has caused a 7,000-square-kilometer cone of depression that has in turn caused more than 1,400 wells to dry up. This same phenomenon has occurred in Taiyuan, where, because of the long reliance on underground water supplies, a cone of depression 100 meters deep has formed and has been expanding at a rate of 2 meters a year, causing land around it to sink.

45. Perhaps the most frightening aspect of this is the sudden and drastic increase in the amount of carcinogens and other harmful substances, which are known to cause mutations in living things, in Shanxi's rivers. Humans may be able to cover their noses and turn away from the river, but the fish have nowhere to run. The population around the Fen River has grown tremendously, and both the Sushui and Mang rivers have seen their fish and crayfish, along with the better-known Yellow River carp, virtually disappear.

46. The other major provincial river, the Wei, has made its own pollution contribution to the Yellow River, something on the order of 600 million tons of wastewater, with significant other amounts coming from the Sushui, Jing, Luo, and, in the upper reaches, the Huang.

47. The first recorded floods occurred as early as 602 B.C. when the great dikes at Guxukou in Ling county completely collapsed, causing the river to change its course from a northerly direction to a more easterly one. Then, 2,500 years later, in 1938, the Nationalist government engineered a massive flood that led to the last change in the river's course so far. When he was threatened by the advancing Japanese army in central China, Chiang K'ai-shek ordered dikes along the southern side at Huayuankou blown up.

48. Scientific studies showed that at least 21 billion cubic meters of Yellow River water were needed to move 1.6 billion tons of sediment to the sea. As the river began to dry up more often, a vast amount of sediment remained on the riverbed. According to some calculations, a billion tons of sediment were left behind in the suspended river in 1997 alone.

The shift in the places where the deposits are left is even more vexing. From 1950 to 1960 conditions in the river were relatively normal: Water flow was adequate and sediment was left as it had been for centuries. In the main bed on the lower reaches only 22.7 percent was left, with the remaining 77.3 percent being piled up on the banks or washed out to sea.

By 1960, with the building of the massive Three Gate Gorge Dam upriver, things had begun to change. Accumulation of sediment on the main bed increased significantly, to 33.5 percent. By the mid-1980s, as the river began to dry up more frequently, large amounts of sediment were now being deposited on the main course—by some accounts as much as 85 percent—making the situation much more serious.

49. In 1958 there were 22,300 cubic meters of water per second flowing across the flood zone, but, fortunately, not one dike was breached and almost no floodwaters were diverted into the drainage area. In the 1980s water flow was measured at 8,000 cubic meters per second and again the major flood barriers were not breached.

In August 1996 water rushed down from the upper reaches at 6,200 cubic meters per second, spilling over into the higher riverbed area at Heze in Shandong province and inundating a 150-square-kilometer section of the Yellow River valley and 298 villages.

In 1998 the Dongming section of the river had a flow rate of no more than 1,920 cubic meters per second, but it flooded anyway, inundating 15,900 hectares of fertile land.

50. Since 1958 the riverbed has risen 4 meters, making an already suspended river even more so. Currently the riverbed on the lower reaches averages 4–6 meters higher than the surrounding countryside. It is 20 meters higher at the city of Xinxiang, 13 meters at Kaifeng, and 10 meters at Jinan.

51. The section between the villages of Taohuayu and Gao is relatively wide at the widest part, about 24 kilometers between the banks and 3–5 kilometers wide on the river itself. Water flow has diminished and serious sedimentation problems have emerged. The river is wide and shallow, rather like a flat plate that cannot hold much water. The water flow has been described variously as "horizontal," "tilted," and "rolling." Conditions like these can do an enormous amount of damage to the embankments and levees. In future they may cause an overflow, putting an end to the "fifty years without a breach." The narrow sections of the river have their own problems. The section below Aishan county in

Shandong is one example. The river is only 1–3 kilometers at its widest and 275 meters at its narrowest. The gradient is only 1/10,000. If floodwaters were to pour down from the upper reaches into this narrow passage, the section could suffer a major breach.

52. Dongping covers a 632-square-kilometer area and is the most important catch basin on the Yellow River. Its primary role is to pull floodwaters off the Yellow River and hold the flow on the lower reaches to no more than 11,000 cubic meters per second, a rate designed to save the city of Jinan, the Tianjin-Huangpu railway line, the Victory oilfields, and the lives of thousands of people living on both sides of the river.

53. The only fully operational dam on the middle reaches is the Three Gate Gorge Dam, which has a capacity of 35.4 billion cubic meters of water and 3.64 billion cubic meters of sediment. The Xiaolangdi Dam is still under construction. It can hold 4.1 billion cubic meters of floodwater and 7.6 billion cubic meters of water in all. Obviously, in the short run, it offers little help in solving these problems. Other projects, such as the Zikou and Guxian dams (12.57 cubic meters and 16 billion cubic meters, respectively), are still in the planning stage. No one has the slightest idea of when they will come on line. The three tributaries—the Yi, Luo, and Qin—drain areas that are frequently hit by rainstorms. Two major water control projects on these waterways—the Guxian and Luhun reservoirs—have a total storage capacity of only 1 billion cubic meters.

54. The forestry protection program planned for the three major northern regions, including much of the Yellow River area, was highly touted. In spite of its being claimed to be the most ambitious project in the world, state spending for the first three years amounted to a mere 60 million yuan, while the funds that local authorities had promised never materialized. Their excuse was the same as always: China is too poor. Yet when it came to some immense water diversion project, somehow those same local authorities managed to come up with big amounts.

CHAPTER TWO

1. The economic importance of the Yangtze River Valley and areas drained by its many tributaries can be seen from agricultural data: It produces 50 percent of the nation's grain, 40 percent of its cotton and cooking oil, and more than 60 percent of its freshwater fish. Because of the convenient and relatively cheap transportation of this "golden waterway," it proved to be a good place to build a long, densely packed corridor of industrial production centers, which ended up accounting for 40 percent of the nation's GDP. These centers can be found in major cities: Sichuan province's Chengdu, the nearby province-level megacity Chongqing, Hubei province's Wuhan, Hunan province's Changsha, Anhui province's Hefei, Jiangsu province's Nanjing, and the other

province-level megalopolis, Shanghai. These helped spur the development of 98 other large and medium-sized cities and 1,900 townships.

2. Prior to the Sui Dynasty (581–618), according to records, the river flooded few times, but by the time of the Tang Dynasty (618–906) this situation had changed radically, with major floods hitting every eighteen years or so. During the Song (960–1279) and Yuan (1271–1368), major floods came more frequently, about every five or six years; in the Ming (1368–1644) and the Qing (1644–1911), it was every four years. From 1921 to 1949 there were major floods on significant parts of the drainage area every two and a half years.

3. They used 350 million bags of sand or other materials, 79 million square meters of coarse cloth, and 43 million cubic meters of sand and rock to stop the floodwaters on the Yangtze and Songhua, but mostly on the former. This was done by at least 6.7 million local people, soldiers, and police who worked round the clock for days on end to shore up levees and embankments. The August sun was intense and the area muddy and pestilential, and some people died from fatigue, exhaustion, dehydration, or disease while manning the dikes. When some of the embankments gave way, it was not uncommon for some people to risk their lives by jumping in to try to fill the gap with more bags.

4. The first began around 315 B.C., following the defeat of the Western Jin kingdom, when large numbers of people went south. They included five ethnic groups from an area north of the Great Wall. When they got to the Yellow River valley there was constant fighting among sixteen overlapping kingdoms that left the region in complete disarray. Many of the northerners fled and ultimately ended up in the Yangtze River valley. What followed was 300 years of turmoil until China was reunited under the Sui and Tang. During the latter dynasty, the rebellion of An Lushan and Shi Siming in 755 caused a second wave of migration to the Yangtze River basin. A third occurred when the Northern Song Dynasty (960–1126) suffered defeat at the hands of a northern tribe. The Northern Song had a population of more than 20 million and large numbers migrated southward in the face of its collapse.

5. During the dry season, large pieces of land in the river valley were put to agricultural use. To protect crops during the flood season from June to November, dikes were built.

People began making the northern banks of the Jing section higher as early as the Eastern Jin Dynasty (317–387), an area that we now know as the Jing River Dike.

6. From 1400 to the end of the Ming in 1644, the amount of arable land in Hubei province increased from 3.6 percent to 17.2 percent of the total area, while in Hunan it grew from 1.1 to 7.8 percent.

7. The Qing court was committed to population growth, and China's population went from 80 million to more than 200 million between 1713 and 1766. The population increase was even greater in Hunan and Hubei. From the last years of the Kangxi reign (1662–1712) to the beginning of Qianlong (1736), a mere thirty years, the number of people in the two provinces grew from 4 million to 18 million. This was not only a result of increased fertility rates but also of significant immigration increases.

9. It has been estimated that from 1685 to 1784 the amount of arable land in Hunan and Hubei increased from 4.18 million hectares to 5.99 million hectares.

10. In 1729, the seventh year of the Yongzheng reign, three Miao areas were put directly under the administrative control of the Qing government. Six years later another prefecture was added.

11. In July 1860 record rainstorms and floods hit the upper reaches of the Yangtze and ran over into the Jing, where peak flows were measured at more than 100,000 cubic meters per second. Low-lying valleys around the Jing were in grave danger and the relatively weak southern side of the river gave way first. Within minutes of the breach, water levels on the Jing dropped 3 meters as the water headed toward the Dongting Plain, which it covered. The Jing River dike held and the densely populated Jianghan Plain was spared much damage, but Hunan suffered major loss of life and destruction of property. Ten years later, heavy rains on the upper reaches caused massive flooding downstream yet again. At its height, water flow on the Jing was more than 110,000 cubic meters per second. By the time the floods reached Songzi county, there was what locals described as something like a huge explosion as the banks of the Jing were breached, but only on the southern side, while the northern dike managed to hold, as it had a decade before. Not surprisingly, on the Dongting Lake Plain there was tremendous loss of life, and both the western and southern regions were completely submerged.

12. In population control there is the famous case in which a person (Ma Yinchu, a demographer and strong advocate of birth control) was wrongly criticized by Chairman Mao Zedong. After that, China's population went from 400 million in 1949 to the more than 1.3 billion at present. The impact this has had on the nation's water supplies has been, according to one estimate, a threefold reduction in per capita access to water. In the Dongting Lake area, the population went from 2.9 million in 1949 to more than 10 million today.

13. In 1951 the Yangtze River Water Resources Commission called for "a comprehensive plan centering on flood gate construction to divert floodwater from the Jing River section of the Yangtze in conjunction with managing the riverbed and Dongting Lake, and all the lakes that feed

into the middle and lower reaches of the Yangtze." The site chosen for the main retention basin was a low-lying, thinly populated part of Gong'an county in Hubei province. In the spring of 1952 the government got 300,000 laborers from around Gong'an county and work on the Jing River retention basin began. The 919 square kilometer basin was finished in a mere seventy-five days and was described as a "great victory for socialism." Two years later the Yangtze River had the largest flood of the century, but because the floodwaters were diverted into basins and lakes on three separate occasions during the onslaught, the loss of human lives and property in Hunan and Hubei was kept down. One place that was completely spared was the city of Wuhan. Large parts of the countryside were less fortunate.

14. In 1954 the population around the major retention basin was only 170,000. By the time of the great flood of 1998, the number had grown to more than 500,000. Many of these people depended on land that had originally been intended to take in floodwaters, not people.

15. In spite of the fact that waters in that area (Shashi) were 22 centimeters above flood level and should have meant the diversion of water through floodgates, no order was given. That spared the lives and property of folks who had been living for several decades in a thousand-square-kilometer area originally intended for use as a retention basin.

16. It soon raised sediment levels in the Jing River significantly and posed a serious threat to Hubei. This deteriorating situation aggravated the already tense feelings between Hunan and Hubei and the debate about which province should be made to bear the brunt. Hunan's proposal that floodgates be built in three other diversion areas was ignored completely.

17. After the great floods of 1952 and 1954, Dongting Lake lost 435 square kilometers to reclaimed land. In 1958, when the government's disastrous Great Leap Forward and the ensuing madness began, land reclamation in the lake area accelerated to a frenzied pace. In only a few years the lake lost 774 square kilometers to agricultural land and residential development. By the 1960s, low-lying regions around Dongting were reclaimed as part of a new campaign, this time to eliminate schistosomiasis, or snail fever. More than 400 square kilometers of the lake area were turned into arable land.

18. Water flow in the Jing did speed up, resulting in increased scouring of the riverbed and a reduction of sediment. But most of the loosened sediment simply ended up settling to the bottom a short way downstream, on the riverbed just below Chenglingji. This, in turn, blocked the Dongting floodwater outlet and raised the lake significantly, which put nearby residents in immediate danger.

19. Back at the turn of the twentieth century, Sun Yat-sen was already convinced that the Yangtze could only be tamed by a huge dam.

His Nationalist government went to work on this dream from the 1920s to the 1940s, with the help of American engineers, and started the laborious task of scouting out appropriate sites. In 1944, a specialist from the U.S. Bureau of Reclamation, John Savage, was invited to study the Yangtze River. He was impressed by the elegance of the great waterway and its immense volume and announced to all who would listen that its unique natural conditions would make his proposed dam project a masterpiece and the capstone of his career: "If God bestows on me sufficient time to see this project through, my eternal soul shall rest comfortably in the Three Gorges." Fate was not so considerate. Mr. Savage was denied his chance by the 1945–49 civil war between the Nationalists and the Communists. China ground to a halt, as did plans for his "masterpiece."

20. December 26, 1970, was the seventy-sixth birthday of Mao Zedong. The aging chairman abruptly announced: "I agree to build the dam." Four days later work began on the Gezhouba dam. Soon afterward it was discovered that there were huge sand deposits around the base of the dam's proposed construction site, and Zhou Enlai put that project on hold, announcing that "unless this problem is resolved similar afflictions will occur on the much larger Three Gorges project."

21. The completion of the Gezhouba in 1978, just 40 kilometers downstream from the Three Gorges Dam site, increased the scouring effect in the Jing River and near the Dongting Lake outlet. However, the water's speed in the Jing River was greater than at the lake's outlets. This caused a reduction in the riverbed's incline, increasing water flow on the upper reaches and turbulence on the lower reaches. That was not conducive to flood control at Dongting Lake.

22. Floods hit 119 counties and municipal areas, 13,898 people were injured or died, and 15.84 million people were affected. More than 140,000 head of livestock were killed, 1.39 million houses were destroyed, 874,000 hectares of land were seriously damaged, and 15 small reservoirs were completely destroyed. Damage was estimated at more than 2 billion yuan.

23. Ganmei accounted for 31.5 percent of Sichuan's area. It saw timber production between the years 1958 and 1994 grow by more than 21 million cubic meters. In recent years, annual forest production has been around 4.8 million cubic meters. Annual growth has only been 2.9 million cubic meters. Along the Dadu, almost all the forests have been cleared, while those along the Yalong are near their end. The tree cutting moved west to Liangshan, where not only were primeval forests in the high country cut down, but cover along the tributaries and the Yalong were destroyed as well. According to a deputy director of the prefecture, even the steepest parts of mountains and hills were not spared.

24. A study was done of the relation of pine and fir forests on the upper reaches of the Yangtze River to soil preservation. It showed that in the oldest primeval growth, even after a huge rainstorm there was virtually no soil erosion. In the younger growth, meaning twenty-three-year-old trees, there was 47 kilograms of soil erosion per hectare, while among the eight-year-old pines the figure was 75 kilograms. In the youngest growth—the two-year-old pines and firs—the soil erosion rose to 1,100 kilograms per hectare. This latter figure is still well below the level of soil erosion taking place on barren hills.

25. Soil erosion estimates by area are as follows: forests, 6 percent; scrub brush, 11 percent; grasslands, 23 percent; and agricultural land, 60 percent. Agricultural lands per se account for 46 percent of the sediment washed into the river.

26. In Ankang prefecture 70 percent of the arable land was on steep hillsides, 23 percent of that on slopes of more than 35 degrees. In Baxian county the situation was even worse. Arable land there with a slope greater than 25 degrees accounted for 63 percent, while 41.3 percent had a slope greater than 30 degrees. On slopes of more than 45 degrees it was 20 percent.

27. Along the Bailong River (another name for the upper reaches of the Jialing in southern Gansu) from 1976 to 1986 deforestation caused 218 million tons of sediment to build up at Bikou Reservoir, putting 41 percent of it out of use and severely damaging another part. The Gongzui Power Station on the Dadu was designed to withstand 30 million tons of sediment annually. Because of unexpectedly high rates of deforestation, it has had to deal with 100 million tons. In a twenty-year time period, the 50-meter-deep reservoir was reduced to only 20 meters deep. That reduced its capacity from 320 million cubic meters to 85 million. On the Yili, a branch of the Jinsha, a trough reservoir with a capacity of nearly 10 billion cubic meters was built in the late 1950s. By August 1977 it had accumulated 7.8 million cubic meters of sediment and its capacity was estimated at less than 20 percent of the original.

28. By 1953 Aba prefecture had become China's premier large-scale logging area, with western Sichuan as its center of operations. In only thirty years' time, 60 million cubic meters of timber were cut, affecting an estimated 200 million cubic meters of forest.

29. According to Chen Qingheng, of the Mao County Biological Research Station, the size of the dry area on the upper reaches of the Min went from 13,300 hectares in the 1980s to 33,300 hectares at present.

30. In the 1930s the river had 17.4 billion cubic meters of water a year; by the 1950s that figure had dropped to 15.7 billion, and in the 1980s to 14.2 billion.

31. The lowest water levels are in February. In the 1930s, they were 161 cubic meters per second; in the 1950s, 143; in the 1980s, 118; and by the end of the 1990s, only 60 cubic meters per second.

32. In 1957 about 460,000 hectares of Sichuan suffered from floods every year. But one result of the cutting of trees during the Great Leap Forward (1958–60) was that floods hit 2.8 million hectares, showing the widespread destructiveness of that campaign. In the years that followed, the flooded areas kept growing annually so that by 1992, more than 5 million hectares of Sichuan were affected. That was equal to 80 percent of the province's entire amount of arable land.

33. In Wudu county, the local meteorology bureau reports that from 1944 to 1950 annual rainfall was 514.2 millimeters. In the 1950s it was 480 millimeters, in the 1960s 471 millimeters, in the 1970s 464 millimeters, and in the early 1990s 439 millimeters. By 1997 it was down to 271 millimeters.

34. A vice governor of Qinghai province told the local media in 1998 that 67 percent of the area around the Yangtze's headwaters had serious soil erosion problems and sent more than 13 million tons of sediment into the river annually. Between 1970 and 1990 the region's glaciers retreated an astonishing 500 meters, and 3.8 million hectares of grassland subsequently became high desert.

35. Huitong county, also on the upper reaches of the Yuan, was known throughout China for its dense forests and timber resources. Forestry Ministry studies have shown that in 1959 its forests covered 17 million cubic meters, but by 1964 that had dropped to 12.97 million cubic meters; by 1973 it was 10.47 million, and by 1980 it had shrunk to only 8 million. That was a 53 percent shrinkage in only twenty years.

36. In 1950 the total amount of forest coverage was 20 million cubic meters. After 26 years that figure had shrunk to 8.5 million cubic meters. As forests disappeared, arable land on steep inclines increased from 20,000 to 58,000 hectares. The area with the worst soil erosion accounted for 41.5 percent. That meant a greater loss of forests. Finally, this became one of the poorest counties in all the mountainous parts of Hunan.

37. By the early 1950s, erosion had affected 11,000 square kilometers. By 1964 the significantly eroded area had grown to 18,000 square kilometers; by 1983 it was 34,000; and by 1987 it was 46,000, or 42 percent of all the mountainous and hilly areas.

38. With such an enormous amount of silt, why are the sediment levels in Poyang lower than those in Dongting, especially since overall Hunan suffers less from soil erosion than does Jiangxi? This anomaly arises from the fact that although Dongting gets heavily silted Yangtze water at several floodgates, which raises its bed 3.5 centimeters annually, in Poyang's case, the shoe is on the other foot. Rather than taking in water from the Yangtze, Poyang actually feeds heavily silted water from

Jiangxi's five major rivers into it. The Hanpokou acts as a conduit, and the Yangtze gets a bad deal by being on the receiving end of Jiangxi's eroded soil. Poyang simply shifts the silt burden of five rivers onto the bed of the Yangtze in its lower reaches.

39. The stocks of fish virtually disappeared almost overnight. But it's not only the fish that have suffered. In 1994 hundreds of people in Baojing county who bathed in the river were taken to the hospital suffering from skin ailments and internal problems after ingesting toxic chemicals. The external effects appeared to wear off, but a more insidious effect began to appear. Cancer rates in the eight towns located on the river's lower reaches started rising every year. Cattle and sheep were hit as well. After drinking from the poisonous river they would cough up a sickening white foam and then drop dead.

40. Concentrations of ammonia, nitrogen, and other chemicals rose drastically while permanganate salts, phenol, and nitric acid were found at category IV pollution levels.

CHAPTER THREE

1. With the highest amount of snowfall annually in all of China, Xinjiang has an annual precipitation of 3.6 billion cubic meters. In the northern part of Xinjiang the annual precipitation is 3.89 billion cubic meters, in the south it is 3.86 billion cubic meters. Put together, they exceed the 5.8 billion cubic meters of annual flow of the entire Yellow River.

2. The river's drainage area is 970,000 square kilometers, or 60 percent of Xinjiang's entire area. Numerous tributaries feed the Tarim. They include the Aksu, Kara Kash, Yarkand, Khotan, Kongqi, Muzart, and many minor waterways. When the Yarkand River (the 970-kilometer-long stream at the head of the Tarim that rises out of the Karakoram mountains in nearby Kashmir) is added, these two waterways measure 2,137 kilometers in length.

3. Bachu county, which is located along the Tarim's upper reaches, devoted virtually all its resources for the past several years to cotton production, a crop with a profligate demand for water that is far more appropriate to water-rich regions such as the United States S'outh. When this lusty demand caused the local Kashgar River to virtually dry up, county authorities authorized the drilling of more than 2,000 water wells. This was completed in only three years' time. This valuable underground source gave the county a 1998 cotton output of more than a million dan (1 million hectoliters).

4. The average number of days during which snow remains on the ground has increased by 8.9, and during the spring and early summer

the snow stays 1.6 days longer before it melts. The average annual depth of the snow has increased by 2 centimeters.

5. Half a century ago the Hei River flowed year round with 1.2 to 1.3 billion cubic meters of water. In the 1960s it had dropped to 1.1 billion, and in the 1970s to 800 million cubic meters. In the 1980s it was down to 500 million, and by 1992 it was a mere 18.5 million cubic meters. The river's dry spells lasted as long as 200 days.

6. The winds swept across the corridor's farmland taking, on average, 1,500 cubic meters of topsoil per hectare. When the winds subsided, the sand that covered the land was 1.5 meters deep, and both electricity and water supplies were cut off. In local irrigation canals, the water was blocked by sand and 3,733 hectares of commercial forests were lost, including 9,000 trees that had been uprooted by the winds. In addition, a vast number of vegetable plots were obliterated and 32,000 head of sheep perished, along with several thousand head of other livestock. In the city of Jinchang, a new metallurgy center in the Hexi Corridor, the winds blew 300,000 cubic meters of nonferrous dust and 166,000 cubic meters of coal dust into the sky, causing the entire city to go from day to night in a dark red and black cloud. People could not see their hands in front of their faces.

7. In Inner Mongolia, the area around the Mu Us (Ordos) Desert was a perfect example of the lush grasslands that characterized this part of China. But as early as the Ming Dynasty (1344–1644) this began to change as the result of a large-scale frontier construction campaign that involved, among other things, an immense land reclamation effort. Dynastic policies encouraged it, and the population along the Great Wall (which was a central part of the Ming frontier defense, mostly against attacks by Mongols) began to increase at an alarming rate. That led to overgrazing and a dramatic increase in the cutting of grass as a cheap source of fuel. Again, almost instantly, the once rich grasslands disappeared without a trace. In their place there appeared more yellow sands that were incapable of sustaining vegetation of any sort. To this day, the ecological damage started by the vast Ming project has yet to be repaired.

CHAPTER FOUR

1. The Nen is 1,370 kilometers long. It starts at a place where the Greater and Lesser Hinggan mountain ranges come together in the eastern part of what was referred to as Manchuria. The Nen is a tributary of the Songhua (Sungari), which is itself the largest tributary of the Amur, and flows directly south to the flat plains of Jilin province. The rich, dense forests around the upper reaches consist of pine, larch, and fir

and extend all the way to the eastern face of the Greater Hinggan range, a large area of wetlands, salt marshes, and brackish lakes.

2. Another assault on the area's forests came from outside in 1896 when the Qing government gave the Russians permission to build the eastern leg of the Trans-Siberian railway via Qiqihar and Harbin and on to Vladivostok. A large amount of timber fell under the axe during the laying of the railway line and went to other development projects along the right-of-way. After the Sino-Japanese conflict in 1895, the Japanese government forced the Qing court to establish the Yalu River Timber Company, which over the course of the next twenty-five years cut down all trees in a 100-kilometer-wide zone along the Yalu's western bank. This massive project continued after the Japanese invaded the area in the 1930s, when there was heavy cutting in both the Greater and Lesser Hinggan Mountains. By the end of World War II it was estimated that the Japanese had taken more than 100 million cubic feet of lumber from the area, leaving the northeast in desperate need of reforestation.

3. In 1949 there were enough large trees that the total was 1.96 billion cubic meters; today the figure is 1.47 billion. In terms of mature trees, the numbers have dropped from 1.45 billion cubic meters to 740 million.

4. The number of Korean pines is down 32.8 percent; camphor pines are down 37.3 percent; and northeast China ash, yellow pineapple, and several others are down by 41 percent.

5. The impact could be seen in places such as the Yichun Forest District, where over the past fifty years the growth of Korean pine and larch was on average a mere 1.5 to 3.5 cubic meters per hectare, whereas considering the amount of underground water as well as heat and soil, the growth should have been around 8 cubic meters per hectare.

6. The swamps are a buffer between the surface and underground water. The water permeates the decaying or dead organic matter of the swamp and helps protect vital groundwater. The swamp and their soils have an enormous water storage capacity. The soil provides the base for the reeds, grasses, and other vegetation. Swamps are referred to as "biological reservoirs." The water capacity of the swamps on the Three Rivers Plain, which is the northeast's largest wetland, is calculated at 3.4 billion cubic meters. That means the swamps were an indispensable link in the ecological chain. A more tangible role was that of the buffer in the event of flooding. The swamps that formed alongside rivers could drastically reduce peak water flows and the impact of the floods. In the Bielahong River valley, a part of the Three Rivers Plain, studies have shown a natural runoff index of .647, meaning water retention ability equal to that of a major forest. The swamps are also important in the impact they have on the climate of their area. The constant cycle of rehydration, evaporation, and dehydration and the decay of organic

matter put a substantial amount of the water into the atmosphere. The Bielahong is an example. Because it is surrounded by swamps, average annual evaporation accounts for 79 percent of its water loss, so the average flow amounts to only 21 percent of its total water volume.

7. Heilongjiang has more than 3.3 million hectares of wetlands and swamps (accounting for 10.8 percent of China's total); in Jilin there are 270,000 hectares; and in Liaoning, 130,000.

8. Shenyang was the industrial masterpiece and capital of Liaoning province. It is now one of the ten most polluted cities in the world. In Jilin province, according to a recent study, the levels of suspended particles in the air of the city of Jilin are eight times higher than the WHO's recommended limit, making it the second most polluted city in the world (the top spot goes to Lanzhou in Gansu province). Then there is the infamous Benxi in Liaoning province, just to the south of Shenyang, where the smog and soot are so pervasive that even spy satellite photographs cannot penetrate.

9. Of the 17,808 kilometers of rivers draining the region, only 160 kilometers have water at level I; the water in 2,984 kilometers (16.8 percent of the total) is at level II; in 3,967 kilometers (22.3 percent) it is at level III, meaning heavy treatment is required for human use; on 6,624 kilometers (37.2 percent) it is at level IV (suitable only for factories or irrigation); and on 4,074 kilometers (22.9 percent) it is at level V (raw sewage) or worse.

10. The southern branch of the river receives the greatest amount of wastewater — 1.2 billion tons annually — or 33.6 percent of the wastewater in the valley. Second is the Nen, with 1.1 billion tons of wastewater annually, or 31.5 percent of the total. The main Songhua takes in 1 billion tons, 28.9 percent of the total. In the Mudan, there is 139 million tons; in the Hulan, 6 million tons; in the Tangwan, 2.5 million; in the A'shi, 2 million tons; and in the Lalin, 1.5 million tons.

11. The most polluted tributary, the Taizi, contains 910 million tons of wastewater. The Hun, which flows through Shenyang, is also seriously polluted, with 742 million tons of waste annually. The main Liao River gets 267 million tons, while the Western Liao gets 122 million and the Eastern Liao 13.5 million tons.

12. In Shenyang, funneling now affects 296 square kilometers of land. The city has drawn its aquifer down from 1.15 billion tons of water to 382 million tons. The water table has fallen from 3.87 meters below the surface in 1949 to 21.5 meters today. In some places it is as low as 40 meters. The same thing has happened in Liaoyang and Fushun, where funneling has hit 320 square kilometers and the water table in the worst areas is down to 24 meters. Even in the parts of Harbin that rely less on underground water, the cone has still managed to cover 500 square kilometers and water levels have fallen 28 meters. In the Daqing oilfield

in the western part of the Songnen Plain, funneling covers 5,000 square kilometers and water levels at its center fell 55 meters.

13. This is evident on one tributary of the Nen, the Huolin, which flows through various parts of Inner Mongolia. In recent years local people have gone in for land reclamation blindly, all in the name of increased grain production. As is usual in these cases, the surrounding grasslands have been overturned, along with the valuable soil. Rainfall here in 1998 was 170 millimeters, which was not much more than in 1995. But in 1995 the river's flow was only a thousand cubic meters per second, and in 1998 it was 3,400 cubic meters a second. An immense amount of floodwater hit towns and villages. The Greater Hinggan forests were hit with a number of mudslides that did a considerable amount of damage.

14. According to the Hydrological Bureau of northeast China, dikes along major rivers will be consolidated so that those on the mainstream of the Nen and Songhua rivers will be able to withstand a 300-year flood. The dikes around Qiqihar, Daqing, and Jiamusi, accordingly, will sustain a 200-year flood, while other cities should reach a 100-year standard and townships a 50-year standard. Finally, rural dikes should reach a 30- to 50-year standard.

CHAPTER FIVE

1. All three rivers contributed to the vast plain, giving it the alternate name of Yellow-Huai-Hai Plain. But the real contribution to the plain's development, the source of deposits, was clearly the Yellow River. The silt levels of the Huai and Hai are much smaller.

2. In 1981, the year of the worst drought, the two reservoirs had only 510 million cubic meters of water left, out of a capacity of 6 billion cubic meters. By June, the order was given to halt irrigation with reservoir water, a move that ended up causing severe damage to more than 2 million hectares of arable land.

3. There were demands placed on the Luan by other cities such as Tangshan, which was hit by a massive earthquake in 1976. The economic development that followed that disaster went on into the 1980s and depended heavily on the river. But the Luan's supplies were limited, so naturally Tangshan had to tap into underground supplies. By the late 1980s the amount of funneling under the city was so extensive that a sudden subsidence of 7 meters left 3,161 square meters of new construction area worthless.

4. The area around Baiyangdian in Hebei province has more than 2 million people and a large number of paper mills, tanneries, and small-scale chemical plants. Their wastewater is dumped untreated into nearby streams and rivers that ultimately make their way into the lake. By far

the largest source of pollution is the city of Baoding, a major industrial center in Hebei, which produces 260,000 tons of wastewater a day.

5. These include Beijing's Huitong River, which is nothing more than a giant sewage disposal for the city, Baoding's Fu River, Shijiazhuang's Jiao River, Datong's Yu River, Cangzhou's Changlangqu River, Xingtai's Niuwei River, and Tangshan's Dou River. All of these carry huge amounts of waste into the lower reaches of the Hai system. The situation has become so bad that some rivers have reached the limit of their ability to absorb waste and sewage. These include the Wei, Nanyun, Yang, Zhang, and the section of the Luan River above Panjiakou Reservoir.

6. There are seventy-one reinforced-concrete sluice gates with serious deterioration problems on the Nanyun and Xuanhui rivers near Cangzhou in Hebei province, while the lift pumps and accessories at five pump stations on the Long River at Langfang in Hebei are so badly damaged that they have to be replaced as often as once a year.

7. The pH of the river water has been as high as 8.6 and its anoxia rate 22.4 times the normal level. Also exceeding acceptable levels are the petroleum content (7.2 times), potassium permanganate (17.5 times), phenol (9.8 times), bacteria content (6.3 times), and E. coli content (6.7 times). Mercury, cyanide, chemical fertilizers, and other toxins have also been detected in the water.

8. A 1990 survey in the southeastern suburbs of Beijing found 100 hectares of land with serious mercury pollution. In the highest instance it was 4.55 milligrams per kilogram of soil. The coarse rice grown in this area was found to contain unacceptably high levels of mercury.

9. The western line, which will link the headwater of Yangtze to that of the Yellow River, was postponed because of its technical difficulties. The entire western line would traverse the major valleys and high mountain arteries of the Hengduan range in Yunnan and the Qinghai-Tibetan plateau, where fragile ecological conditions would undoubtedly be harmed by the construction and operation of such a large-scale project. Considering that the Hengduan mountains and Qinghai-Tibetan Plateau were created by the collision of the South Asian and Eurasian tectonic plates, geological formations in this area are extremely unstable, as indicated by the numerous earthquake belts and fissures dotting the region. To say the least, these are not the best place to build such a large-scale project conveying billions of cubic meters of water.

10. In the Hai River valley, like many other places in China, soil erosion is quite serious: In the early 1980s 110,000 square kilometers, covering nearly 60 percent of mountainous areas, was affected, with total topsoil erosion of more than 400 million tons. The impact has been quite dramatic, as the area has witnessed not only increasingly frequent

droughts but also sedimentation blockage of reservoirs and flood diversion areas along with elevated riverbeds. Data from the region indicate that total sedimentation in twenty-four large-scale reservoirs is as high as 170 million cubic meters, or 10.5 percent of total reservoir capacity. Take, for instance, the Guanhe reservoir, where the sedimentation problem is probably the most serious in that 46.8 percent of its water retention capacity has been compromised by sedimentation buildup. No better is the situation in the Guanting reservoir, just outside Beijing, where more than 600 million cubic meters of water storage capacity has been lost to sedimentation build-up.

11. One such case involved the Foziling reservoir, located in Huoshan county, the first large-scale water retention facility built in a mountain valley in China. Soon after its completion in 1954, many cracks were found in the primary structure of the dam. Then in 1965 and 1969, despite two efforts at reinforcing the dam, the problem was still not resolved. Finally, on October 21, 1993, measurements at the dam site indicated that the structure was shifting at a rate far beyond the allowable limit. Today, Foziling is still considered a "sick dam" whose water retention capacity is strictly limited and which, if threatened by flooding, could collapse at any time. Similar problems have been found at the Meishan dam, where inspectors discovered more than 1,000 cracks on its arched structure; significant water leakage is still occurring. Currently, critical parts of the dam's intake and discharge systems are in need of repair, and overall the basic structure is not sound. And at the Xianghongdian dam, which lacks any sluice gates, cracks have already been found on the left side of its frontal facade.

12. Of the wastes released into the river there are 188 thousand tons of ammonia, 122 thousand tons of suspended particles, 18.5 tons of various heavy metals, 92.4 tons of evaporative phenol, 9.8 tons of cyanide, and 4.8 tons of arsenic.

13. Dingxiang Township, for instance, located along the Ying River, has 800 such factories that, in order to soften the leather, employ a metallic chemical element known as red vanadium, which is extremely poisonous and leads to the generation of large amounts of liquid wastes containing chromium. If more than 0.5 grams of chromium is absorbed into the human body, it can lead to cancer and death.

14. According to statistics, from 246 B.C. to 1949 and on to the year 2000, on average each century has witnessed twenty-seven major floods along with thirty-five significant droughts. But once the Yellow River skirted the Huai River and blocked its entry into the sea, sometime in the twelfth century, the number of average floods and droughts hitting the valley per century increased dramatically, to ninety-four and ninety-nine, respectively. In fact, the Huai River has gained a national reputation as a region where "huge rainstorms produce big floods, small rainstorms cause small floods, and the lack of rainstorms causes drought."

CHAPTER SIX

1. The "new China," founded in 1949, did not put an end to lake reclamation in the Jiangnan region. In fact, it led to a new wave of development. Over the next four decades more than 530 square kilometers of lakes were reclaimed in the Lake Tai region, including 160 square kilometers of Lake Tai itself, with a net reduction in its 600 million cubic meters of water, leaving it vulnerable to flooding.

2. Hangzhou was the capital of the ancient kingdom of Wuyue and, later, of the Southern Song Dynasty. It has long been known as one of China's six most prominent old cities. It is located on the lower reaches of the Qiantang River and at the southern end of the Beijing-Hangzhou Grand Canal. The main source of its prominence in Chinese history is West Lake, in the center of the city.

3. The overuse of groundwater has led to subsidence across the region. In Shanghai, which used large amounts of underground water from 1949 to 1963, the water table fell to more than 30 meters below ground. The largest drop, in one part of the city, was 2.6 meters. This had an impact on high-rises, with cracks appearing on building facades and pipes rupturing, while the foundations of bridges and surfaces of roads and even airport runways developed large cracks and other problems. The city center is 2.1 meters below sea level. Similar subsidence has occurred in other cities in the region.

4. The three rivers are the Hai, Huai, and Liao. The three lakes are Lake Tai; Dianchi, in Yunnan province; and Chao, in Anhui province.

5. In the city of Jiaxing in Zhejiang province, which lies next to the lake, the incidence of intestinal infections and water-borne diseases such as hepatitis A, diarrhea, and typhoid fever is quite high, about 70 percent higher than in the rest of the province.

6. The first assault on the region began as long ago as the Western Jin Dynasty (A.D. 265–316) when, as the Han and sixteen separate ethnic minorities took turns running the regime, many people in northern China were driven south of the Yangtze by nomadic invaders. They reclaimed the virgin area with their advanced tools and agricultural techniques, and forest cover on the southern plains was quickly destroyed.

The second large-scale destruction of Zhejiang and Fujian occurred during the Song Dynasty (960–1279). The population of China was already approaching 100 million, and almost all its plains had been put under the plow. During the Northern Song, mountainous areas were opened up for peasant settlements. From 1080 to 1223 the number of households in this area grew by 600,000. In that time, the area's households went from a mere 6.67 percent of China's total to 12.62 percent.

One of the most destructive tools in this devastation was the increased use of corn and other crops on hillsides by people who became known as the "shack people" (*peng min*). During the Kangxi and Qianlong eras of the Qing Dynasty, the shack people got such large profits from corn that farmers from all over rushed to the mountains to grow even more of this high-yield crop. The impact on population growth was enormous. From 1713 to 1791, Zhejiang's population went from 2.7 million to 23 million.

7. In 1936, a study of the headwaters of the rivers and streams that fed Lake Tai found that the degradation caused by the overuse of upland areas by the "shack people" was causing deforestation. Soil erosion was so serious in many places that what had once been cultivatable soil had turned as hard as stone; large forest areas virtually disappeared, and mountain streams dried up. Floods were uncontrollable because of a lack of tree cover. The consequences were dire.

8. Estimates are that only twenty or so still exist, living virtually alone in the few remaining forest "islands." They are not expected to last very long.

9. The timber industry was one example. In the last fifty years Zhejiang has provided the state with more than 21 billion cubic meters of lumber and 700 million bamboo poles. But that's not all. The state accounts for less than a fourth of the total consumed. That means that the amount of lumber produced during that time was more than the amount of timber that has been left standing. Fujian is currently China's third-largest lumber producer, according to the state plan. By 1980 it had provided 80 million cubic meters of lumber and 300 million bamboo poles.

10. Mountain streams in Fujian have had far more dry periods during spring and summer compared to the 1950s. The increasing frequency of droughts and floods makes Fujian not much different from the rest of China. In the 1990s, its mountainous regions were hit with floods, and in 1998 it had its biggest flood ever. The mountainous regions of Zhejiang province are not much better, with 1.33 million hectares having been declared barren.

11. Shuikou Reservoir, which was built in 1993, has suffered from increasing amounts of pollution and has also had a negative effect on the ecology. It has caused tremors in the earth that have affected irrigation networks by reversing the flow of water and increasing seawater levels upstream. It has also accelerated the speed of soil and water erosion on streams above it. The reservoir contributed to the collapse of riverbanks nearby and is said to have been a major factor in the structural failure at Liberation Bridge over the Min. Qiantang River Reservoir, which is equal in size to more than 3,000 West Lakes, has seen an immense increase in the amount of organic nutrients and algae.

12. In fact, the Pearl River (South Pearl) refers only to that part that flows from the city of Guangzhou (Canton) to the sea, a mere 96 kilometers. It got its name from Pearl Island, located in the center of the stream near the sea. The three tributaries—the West, North, and East rivers—flow independent of each other until they converge in the Pearl River estuary and flow into the sea as one.

13 Guangxi has made some progress since 1949 in restoring its forests, but at the same time its population has gone from 18 million in 1950 to more than 50 million today. Its desire for economic advancement has meant turning to the mountains for prized goods when other forms of capital, technology, or natural resources are unavailable. Estimates are that 23 million cubic meters of trees were added to the growth in Guangxi between 1980 and 1985, while 28 million cubic meters were consumed. Before 1987 67 million hectares of forests were destroyed each year; there were some constraints on logging and tree cutting imposed in subsequent years, but not everywhere. After so many years of destruction, the quality of timber in Guangxi has declined dramatically, on average yielding less than half the national average, which is itself quite low when compared to other parts of the world.

CHAPTER SEVEN

1. Each year the mudslides destroy a considerable number of railways, roads, bridges, and tunnels. They've hit the 81-kilometer-long Dongchuan Railway line, which, in spite of its immense construction cost, was rendered virtually useless after only twenty years of operation because of mudslides. This was the first case in China of a railway being completely destroyed by mud. The rivers of mud have wreaked havoc on the reservoirs and streams and have blocked several, not only causing economic damage but also making flood conditions worse.

2. Forest cover in the Dian Lake region is currently about 21 percent, but that's a 16.4 percent reduction from 1951. In the past forty years 56 million cubic meters of sediment have accumulated in the lake, and it has become quite shallow. That has in turn affected its ability to regulate floodwaters and has led to intense water conflicts in the valley. At present, the amount of per capita water availability in the valley is only 310 cubic meters, or an eighth of the national average. As sediment built up in the lake, farmers reclaimed more land for agriculture.

In the past few decades, lake land has been used for all sorts of additional purposes like housing and tourism, causing the area to shrink even more. In 1985 there was an effort to put a halt to unplanned lake reclamation, and Yunnan authorities promulgated a "Protect Dian Lake Order" (Dianzhi baohu tiaoli) at a time when the body of water had already shrunk to a mere 306 square kilometers. Local people virtually ignored the regulations and continued their frenzied reclamation habits, so by 1995 it had shrunk to 292 square kilometers.

3. The city of Kunming relies on lake water for the chemical, food processing, paper milling, textile, and coke refining industries. In 1992 more than 300 million cubic meters of industrial wastewater was released into the lake; in 1995 the discharge amounted to nearly 500 million cubic meters. Then there was the destructive agriculture in the lake valley, especially when farmers joined their brethren everywhere else in China in falling in love with chemical fertilizers, whose residue inevitably ended up in the lake. The area's population grew substantially, to 2 million, and human waste from the improved lifestyles found its way into the lake. In 1992 there was 94 million cubic meters of residential wastewater discharge into it; in 1995 it grew to 135 million cubic meters; and in 1999 it topped out at 185 million cubic meters.

BIBLIOGRAPHY

PERIODICALS AND NEWSPAPERS (CHINESE)

Beijing Economic Report (Beijing jingji bao).
Beijing Water Resources (Beijing shuili).
Chengdu Water Resources (Chengdu shuili).
China Agricultural History (Zhongguo nongshi).
China Economic Daily (Zhongguo jingji ribao).
China Environmental News (Zhongguo huanjing bao).
China Industrial and Commercial Times (Zhonghua gongshang shibao).
China Meteorological Daily (Zhongguo qixiang bao).
China Water Resources News (Zhongguo shuili bao).
China Youth Daily (Zhongguo qingnian bao).
China's Three Gorges Project (Zhongguo sanxia jianshe).
Contemporary Pearl River Development (Zhujiang xiandai jianshe).
Economic Daily (Jingji ribao).
Economic Reference News (Jingji cankao bao).
Enlightenment Daily (Guangming ribao).
Environmental Protection (Huanzhing baohu).
Farmers' Daily (Nongmin ribao).
Fresh Water Lakes Science (Hubo kexue).
Fujian Environment (Fujian huanjing).
Geographical News (Dilixue bao).
Geography and National Lands Research (Dilixue yu guotu yanjiu).
Green China Tribune (Zhongguo lüse shibao).
Guangxi Water Resources (Guangxi shuili).
Hai River Water Resources (Haihe shuili).
Harnessing the Huai (Zhi Huai).
Heilongjiang Daily (Heilongjiang ribao).
Hunan Forestry (Hunan senlin).
Hunan Water Resources (Hunan shuili).
Jiangsu Water Resources (Jiangsu shuili).
Jiaqingpeng County Archive.
Legal Daily (Fazhi ribao).
Literary Gazette (Wenhui bao).
Managing China's Environment (Zhongguo huanjing guanli).

Managing the Southwestern Environment (Xinan huangjing zhili).
Neighborhood (Fangyuan).
Northeast Water Resources and Hydro-Power (Dongbei shuili shuidian).
Outlook News Weekly (Liaowang xinwen zhoukan).
Outlook Weekly (Liaowang xinwen zhoukan).
People's Daily (Renmin ribao).
People's Pearl River (Renmin zhujiang).
Pollution and Protection of China's Water and River Systems (Shui xi wuran yu baohu).
Protection of Water Resources (Shui ziyuan baohu
Qinghai Water Resources (Qinghai shuili).
Reader's Digest (Baokan wenzhai).
Regional Studies and Development (Diyu yanjiu yu kaifa)
Science and Technology Daily (Keji ribao).
Sichuan Water Resources (Sichuan shuili).
Southern Weekend (Nanfang zhoumo).
Tibet Daily (Xizang ribao).
Urban Areas Herald (Chengshi daobao).
Voice of China (Huasheng bao).
Water Systems Pollution and Preservation (Shuixi wuran yu baohu).
Worker Daily (Gongren ribao).
Xinhua Daily Wire (Xinhua meiri dianxun).
Xinjiang Daily (Xinjiang ribao).
Yunnan Environmental Sciences (Yunnan huanjing kexue).
Yunnan Geographical Environment Research (Yunnan dili huanjing yanjiu).

BOOKS (CHINESE)

Beijing Weekly, ed., *Three Gorges as a Key Point Project* (Sanxia shuili shuniu) (Beijing: New Star Publishing House, 1991).

Chen Hua, *Environment and Population Growth in the Inner Mongolian Autonomous Region* (Neimenggu zizhiqu renkou fazhan yu huanjing) (Beijing: China Environmental Sciences Publishing House, 1995).

China's Population and Environment–Volume Two (Zhongguo renkou yu huanjing--di erji (Beijing: China Environmental Sciences Publishing House, 1995).

China Statistical Bureau, *1999 Statistical Yearbook* (1999 Zhongguo tongji nianjian), China Statistical Publishing House, 1999).

Department of Pollution Control, State Environmental Protection Bureau, *Countermeasures Employed in Pollution Control in China's Environment* (Zhongguo huanjing wuran kongzhi duice) (Beijing: China Science and Technology Publishing House, 1998).

Destruction of Forest Reserves Resulting From the 1998 Flood (Jiuba hongshui jujiao senlin), (Beijing: China Forestry Publishing House, 1999).

Destructive Effects of the 1998 Flood on Forests (Jiubao hongshui jujiao senlin) (Beijing: China Forestry Publishing House, 1999).

Gansu Province Statistical Bureau, *Population and Environment in China, Volume II* (Zhongguo renkou yu huanjing di'erji) (Beijing: China Environmental Sciences Publishing House, 1995).

Guo Hansheng, Yang Zhibao, Wang Shanbian, *Water Resources: Basic Knowledge and Facts* (Shui ziyuan zhishi wenda) (Zhengzhou: Yellow River Water Resources Publishing House, 1997).

History of Geography, Volume Two: Lessons from the Degradation of Lake Regions in Ancient China (Lishi dili di erji: wo guo gudai hubo de yanfei ji qi jingyan jiaoxun) [n.p., n.d.].

Hong Qingyu, ed., *Flood Control in China: A Series—The Yangtze River* (Zhongguo jianghe fanghong congshu--Chiangjiang *juan*) (Beijing: China Hydropower Publishing House, 1998).

Huai River Water Conservancy Commission, *Series on Flood Control on China's Rivers: Huai River Volume* (Zhongguo jianghe fanghong congshu: Huaihe juan) (Beijing: China Water Resources and Hydro-Power Publishing House, 1996).

Huang Xiguan, Su Fachong, and Mei Anxin, *China's Rivers* (Zhongguo de heliu) (Beijing: Commercial Press, 1996).

Li Xianwen and Guo Kongwen, editors, *One Hundred Questions Regarding the Great Flood of 1998* (Jiuba dahongshui baiwen) (Beijing: Hydropower Publishing House, 1999).

Liang Yijun, *Lumberjacks, Wake Up!* (Famuzhe, xinglai!) (Beijing: China Environmental Sciences Publishing House, 1997).

Lin Yingshuang and Zheng Kaige, eds., *Value of Life's Cradle* (Zhenxi shengming yaolan) (Beijing: China Environmental Sciences Publishing House, 1997).

Liu Xiaoyan, Wu Zhirao, and Yao Chuanjiang, *Harnessing the River With Science and Technology* (Keji zhihe) (Zhengzhou: Yellow River Hydrological Publishing House, 1997).

Long Shengsheng, *Agricultural Geology in the Two Lakes Region During the Qing Dynasty* (Qingdai lianghu nongye dili) (Wuhan: Central China Normal University Publishing House, 1996).

Lü Enlin, *Environmental Management of Southwestern China* (Xinan huanjing zhili) (Kunming: Yunnan Education Publishing House, 1992).

Ma Lihua and Lan Zhiming, "Developmental Projects in the Middle Reaches of the Yarlung Zangbo Benefitting the Tibetan People," in *Series on Tibet* (Xizang congshu) (Beijing: Five Continents Broadcasting Publishing House, 1999).

Ma Zhongliang et al., *China's Changing Forestry* (Zhongguo senlin de bianjian) (Beijing: Forestry Industry Publishing House, 1996).

Ma Yingxian, *A Geography of Xinjiang's Vistas* (Xinjiang lüyou dilixue) (Ürümqi: Xinjiang Handicrafts and Photography Publishing House, 1993).

Measures for Controlling Water Pollution in the Hai River System (Qiaoxiang Haihe liuyu shui wuran fangzhi de jingzhong) [n.p., n.d.].

Research on the Social-Economic History of Contemporary China (Zhongguo jinshi shehuijingjishi yanjiu) [n.p., n.d.].

Song Zhaoshu and Zhang Qinghua, *Future Grand Forests: Forestry in the Twenty-First Century* (Da senlin de wilai: 21 shiji de senlin) (Beijing: Science and Technology Abstracts Publishing House, 1995).

Wang Jingxiang, Yao Jiheng, Niu Ruiyan, *Zhejiang Forestry* (Zhejiang senlin) (Beijing: China Forestry Industry Publishing House, 1993).

Wu Zhonglun, ed., *China's Forests* (Zhongguo senlin), Vol. I (Beijing: China Forestry Industry Publishing House, 1997).

Xing Shaoming, ed., *Jilin Forestry* (Jilin senlin) (Changchun: Jilin Science and Technology Publishing House, 1995).

Xu Gang, *Vicissitudes of Water Flow: Volume on Rivers* (Liushui cangsang:hejiang zhi juan) (Changsha: Hunan Science and Technology Publishing House, 1997).

———, *Ecological Changes Throughout Chinese History* (Zhongguo lishi shang shengtai huanjing zhi bianqian) (Beijing: China Environmental Sciences Publishing House, 1996).

Xu Zhiyang, ed., *Natural Resources and Development in the Yangtze River Delta* (Changjiang sanjiao zhou shuitu ziyuan yu yu fazhan) (Hefei: China Science and Technology Publishing House, 1997).

Yuan Kechang, *Threat to Existence: Pollution* (Shengcun de weixie: wuran) (Nanjing: Jiangsu Science and Technology Publishing House, 1998).

Zhang Haobai, ed., *Fujian Forestry* (Fujian senlin) (Beijing: China Forestry Industry Publishing House, 1993).

Zhu Niuzhuan and Ma Xuehui, *China's Wetlands* (Zhongguo de zhaoze), (Beijing: China Commercial Bookstore, 1996.

Zhu Zhenda and Chen Zhiqing, *Yellow River Dry-Up and Sustainable Development of the Yellow River Basin* (Huanghe duanliu yu liuyu kechixu fazhan) (Beijing: China Environmental Sciences Publishing House, 1997).

Zhu Lanqin, ed., *Three Hundred Questions Concerning the Yellow River* (Huanghe sanbai wen) (Zhengzhou: Yellow River Water Resources Publishing House, 1998).

BOOK (ENGLISH)

Sandra Postel, *Last Oasis: Facing Water Scarcity* (NY: Norton, 1992).

APPENDICES

China's Major Drainage Basins

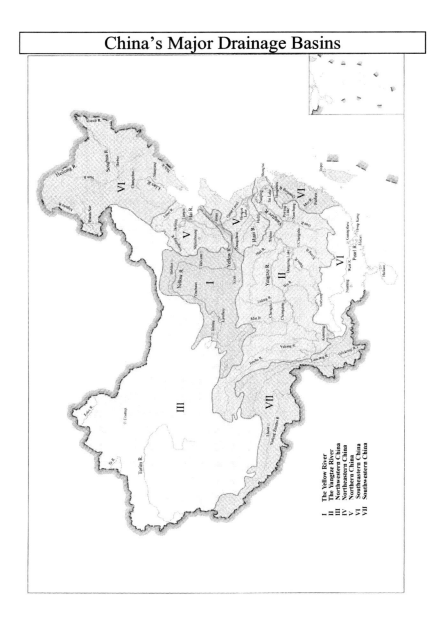

I   The Yellow River
II The Yangtze River
III Northwestern China
IV Northeastern China
V   Northern China
VI Southeastern China
VII Southwestern China

China's Deserts and Land hit by Desertification

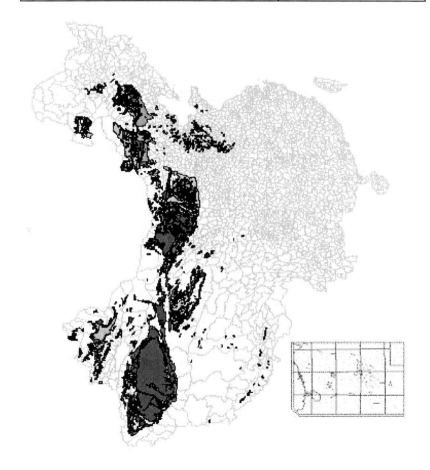

Yellow River Drainage Basin

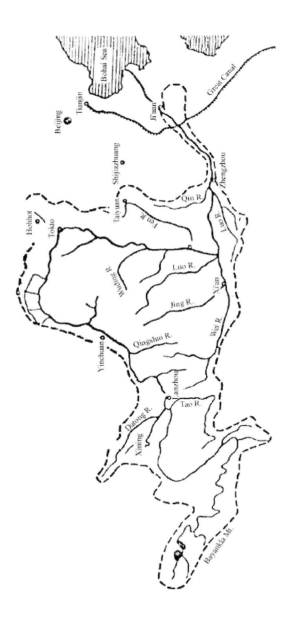

Loess Earth and Rock in China

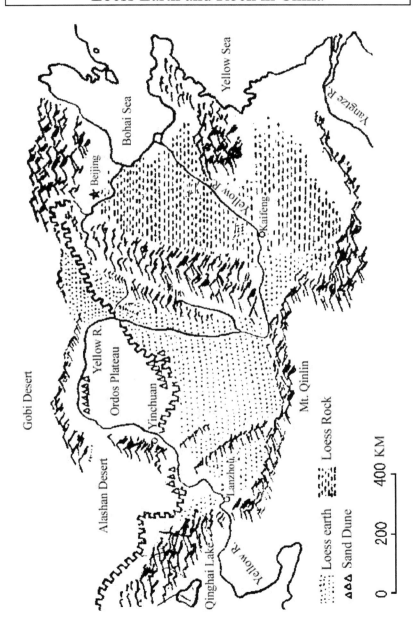

Yangtze River Drainage Basin

Lake Reclamation on the Jiang Plain

Tarim River Basin

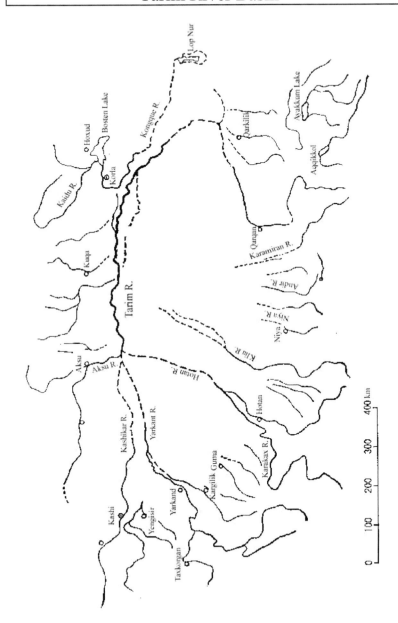

Northeastern Rivers

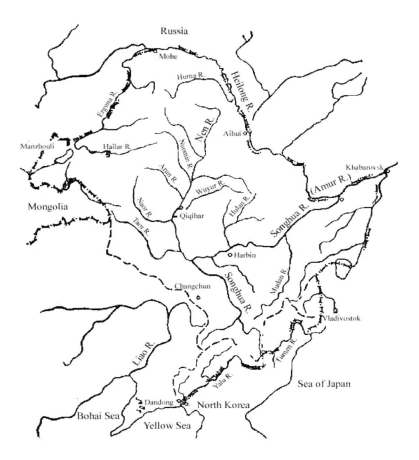

Huai River Drainage Basin

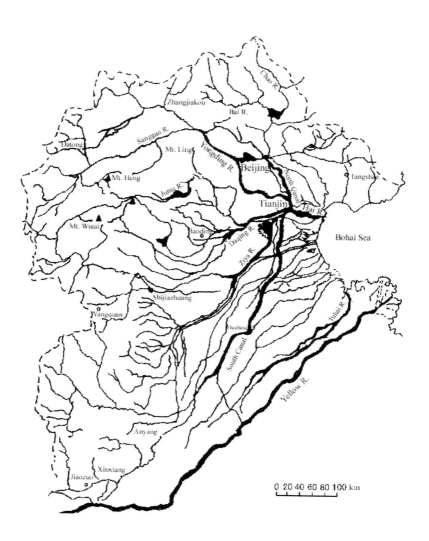

Lake Reclamation in the Ming Dynasty

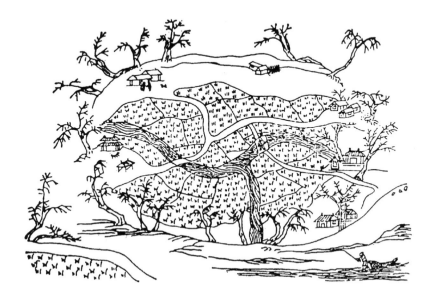

Location of Lake Tai

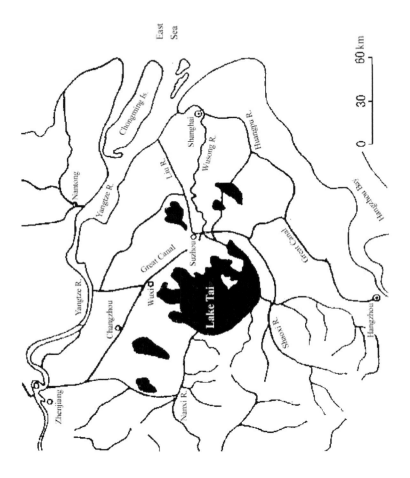

Pearl River Drainage Basin

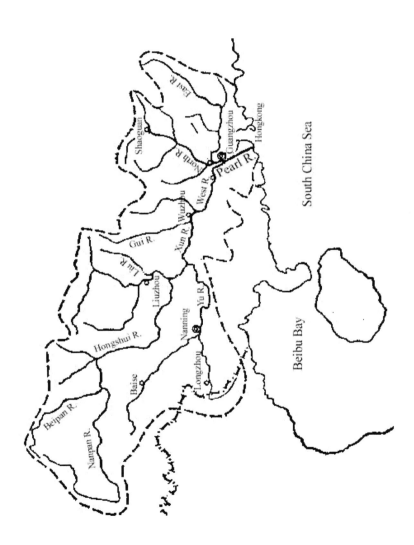

Major Rivers Originating in Southwestern China

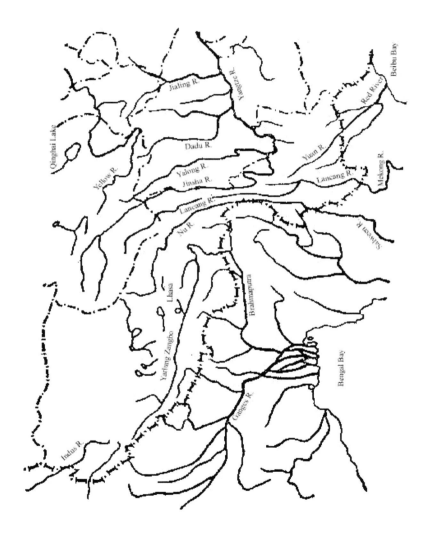

Major Lakes on the Qinghai–Tibet Plateau

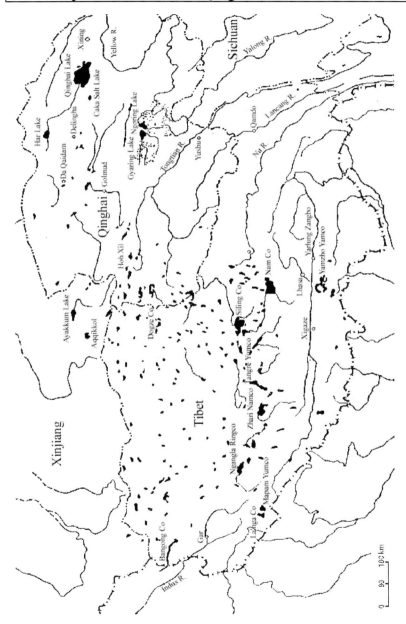

INDEX

CHINA'S WATER CRISIS

Ma Jun, a leading Chinese investigative journalist, worked at the *South China Morning Post* from 1993 to 2000 where he produced his own reports and wrote many feature articles on the Chinese environment, eventually becoming Chief Representative of SCMP.com in Beijing. He is currently an environmental consultant with Sinospere Corporation.

Nancy Yang Liu is a professional translator.

Lawrence R. Sullivan is Associate Professor of Political Science, Adelphi University.

EastBridge

Voices of Asia

Steven I. Levine, Imprint Editor

Voices of Asia presents important books in the social sciences and the humanities by leading contemporary Asian writers in English translation. By making available work which would otherwise be accessible to only a very small number of Western specialists, *Voices* facilitates a two-way flow of knowledge between Asia and the West.

Steven I. Levine teaches Asian history at the University of North Carolina at Chapel Hill and is interim Director of the Carolina Asia Center. He received his AB in Politics from Brandeis University and Ph.D. in Government and Far Eastern Languages from Harvard University. He has taught at the University of Michigan, Duke University, The American University, the Defense Intelligence College, Columbia University, and Merrimack College. He served as a Consultant and Social Scientist at the Rand Corporation; as Director, Center for Slavic, Eurasian and East European Studies, University of North Carolina at Chapel Hill; and as Resident Director, Duke Study in China Program, Bejing and Nanjing.